All About Bees,
Beekeeping
and Honey

All About
Bees,
Beekeeping
and Honey

Walter L. Gojmerac

Professor of Entomology

University of Wisconsin

DRAKE PUBLISHERS INC.

NEW YORK • LONDON

Published in 1977 by
Drake Publishers Inc.
801 Second Avenue
New York, N.Y. 10017

Library of Congress Cataloging in Publication Data

Gojmerac, Walter L
    All about beekeeping, bees, and honey.

    1. Bee culture. I. Title.
SF523.D63   638'1   76-16368
ISBN 0-8473-1310-7

Printed in the United States of America

CONTENTS

All About Bees, Beekeeping and Honey

Preface

THIS BOOK IS DESIGNED to enlighten anyone who is interested in nature's most useful and misunderstood creature, the indispensable honeybee (*Apis mellifera*). Interest in bees is not new. Since recorded history man has observed, studied, transported, admired, and feared bees. He has also robbed their honey and probably will continue to do so for generations to come. Man's interest in and respect for bees increased when it was learned that they were responsible for pollinating at least 50 food crops. While honey was replaced by sugar as the most popular sweetener on a per-capita basis, many people still prefer and buy honey, the natural sweetener. Intensified agricultural practices have reemphasized the need for large numbers of properly managed colonies. They must be available at precise times to ensure pollination of crops such as apples, almonds, alfalfa seed, cucumbers, and many others.

Beekeeping can be an interesting and challenging hobby or part-time business enterprise for many people from all walks of life. When people turned to science for answers to practical and theoretical problems in sociobiology, the bee became an object of study and admiration. How can thousands of individuals live in as small an area as a hive? Do they have a hierarchy of administrators? Their language has been translated, so we know that they have a communication system. Their society is directed in part by chemicals called pheromones. People have spent a lifetime studying the biology, behavior, and activities of bees. This book may not answer all your questions — no single book can. It does summarize some of the latest scientific information on bees, their activities, and their products. If you are not familiar with bees, this book will introduce you to some of the mysteries and secrets of the hive and perhaps motivate you to seek further information about bees, beekeeping, and honey.

The information presented is as technically accurate as possible, based on published literature and on knowledge gained from my personal experience as a biologist, entomologist, and amateur beekeeper. For the nonprofessional the latest technical principles are summarized in an easily understood, logical sequence. Beekeepers are notorious for their strong differences of opinion. It is often said that if 10 beekeepers met to discuss a problem, they would usually arrive at 11 different solutions. If you disagree with some of the statements, do not be upset — you are a normal, progressive beekeeper. For nonbeekeepers about to read this book, it is my hope that it will excite a lifelong interest in the fascinating honeybee.

The Honeybee:

A Unique Creature

MAN'S INTEREST IN BEES is as ancient as recorded history. Why are people so attracted to this creature? Few understand this strange phenomenon. Perhaps man instinctively enjoys eating sweets, and honey first attracted him to the honeybee. The significance of pollination was not recognized until after 1750, and only recently has its true importance been understood. Most amateur beekeepers recognize and appreciate the value of pollination, but something else sparks their interest. Bees are intriguing creatures. Their industry is admired by some, envied by others. Their engineering capability, evident in comb building, is highly respected by structural engineers. Their defensive capability is respected by man and even by some animals. Yes, the honeybee is truly a unique creature.

Classification

In 1758 Linnaeus devised a system of classifying living organisms. Each organism is given two names, genus and species. The honeybee is called *Apis mellifera*. A group of closely related insects belongs to the same genus. Several closely related genera compose a family. In this case all bees, including the bumblebee and the carpenter bee, belong to the same family, Apidae, as the honeybee. An order is a collection of closely related families. Wasps, hornets, yellowjackets, bees, and several lesser-known groups all belong to the same order, Hymenoptera.

Within the species *Apis mellifera* are several races, sometimes referred to as subspecies, which can and will interbreed. Races of bees differ from varieties, breeds, or races of commonly known plants and animals, many of which are man-made selections and crosses. Races of bees developed naturally in isolated geographic regions. Three general races of honeybees are recognized: oriental, European, and African. These general races are further subdivided: for example, within the European race are the Italian, Caucasian, Caroniolian, and dark bees of northern Europe. *Adansonii* is one of several African races. The correct designation for the Italian bee is *Apis mellifera ligustica;* for the dark European, *Apis mellifera mellifera;* and for the African, *Apis mellifera adansonii.* These races are referred to in this book by their common names.

Where and how did the present honeybee develop? Fossil records are not abundant, but it is generally believed that it originated somewhere in southeast Asia. From this area the insect migrated over millions of years to the tropics and to temperate and northern regions of Europe and Asia. Those that moved to the tropical regions had no reason to seek shelter in cavities or hives, nor did they accumulate large stores of nectar and pollen, apparently because it was available all year long. Even today some bees in the tropics of southeast Asia build a single large comb out in the open. According to some scientists communication is not highly developed in this group. The bees that moved to temperate or northern regions met a more hostile climate, in which periods of drought and/or cold winters were followed by periods of luxuriant growth, providing abundant nectar and pollen. Colonies that could capitalize on this phenomenon survived and prospered and it is from this group that the honeybee as we know it to-day undoubtedly developed.

Bees are social creatures: an individual bee cannot exist by itself. Each individual must be a member of a group, a society called a colony. The colony is a true organism. It is alive; it grows; it gathers eats, and stores food; it is capable of reproduction. The colony is normally considered perennial, but if conditions are unsuitable, it will die.

A hive is a man-made structure designed to house the colony. The conventional standardized hive consists of a top cover, inner cover, bottom board, and several chambers, called hive bodies, brood chambers, or supers. Inside each chamber are a number of frames on which bees build vertical combs of wax. The top bar of the frame is designed to suspend the comb within the hive body. The brood chamber, within the hive body, houses the queen, eggs, larvae, and pupae. This area is collectively called the brood nest; larvae and pupae are often called brood. Honey and pollen are stored in specific areas of each frame. The chamber of compartment on top of the brood is called a super or honey super, and honey is naturally stored above and around the sides of the brood nest.

The colony, as a living organism, reproduces itself. If the colony becomes crowded because of a large number of bees, workers begin to build specialized cells, usually on the lower surface of the frame, called swarm cells. They are vertical rather than horizontal, which is the normal position for worker and drone cells. As crowding increases, the queen lays an egg in each cell. After the egg hatches, workers feed the developing larva a specialized diet in order to produce a queen. Before the developing new queen emerges, the old queen and a large percentage of workers usually leave the colony in a swarm. If a nearby cavity or other suitable structure is available, the swarm will enter and set up housekeeping, establishing a new colony. Meanwhile, in the old colony the new queen will emerge. She will fly out to mate and return, and life will continue. Swarming is the natural way in which colonies reproduce.

Early History of Beekeeping

Beekeeping (apiculture) evolved long before the true importance and significance of bees were understood and appreciated. Somewhere between 7000 and 3000 B.C. the ancient Egyptians depicted bees in drawings on tombs. These pictures suggest that they knew the value of smoke in working with or handling bees. In 334 B.C. Aristotle recorded his observations on bees in his treatise *Natural Sciences*. He recognized several races. His writings suggest that he might have had some type of movable combs or possibly an observation hive. In 40 B.C. Virgil described how to make cork hives. He recommended that they be placed in the shade and protected from cattle. He also knew that bees need an adequate supply of honey to survive. Through the centuries other people wrote about bees, but they often mixed observations with imagination without referring to facts, so progress was slow.

In the Reformation a number of monastaries were destroyed, which dealt beekeeping a severe blow. Many monks had been beekeepers. They made hives from pottery, coils of twigs or reeds, straw, or simply a hollow log. These hives protected the colony from the elements. Flight entrances were small, and some hives had a second opening for removing honey. The early beekeepers kept hives

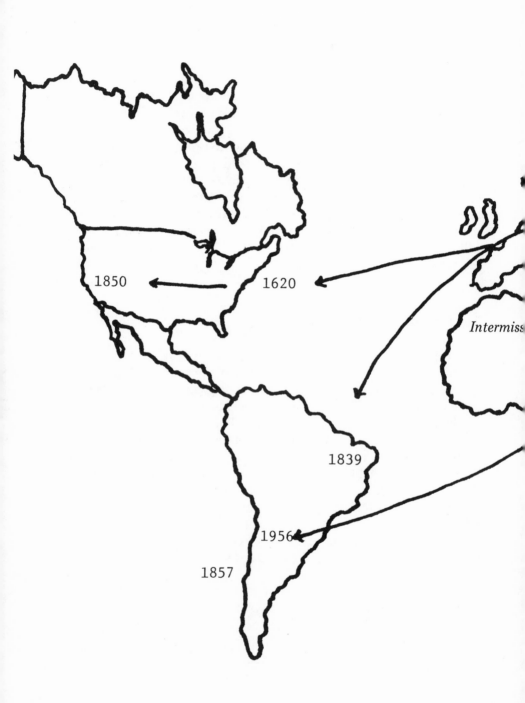

Origin and distribution of bee races.

small and encouraged swarming in order to populate empty hives. At the end of the season the bees were killed, usually with hot water, and honey and wax were recovered.

By the early 17th century the more progressive beekeepers tried to devise ways of recovering honey without killing the bees or destroying the combs and hives. Some united small colonies into large ones, others constructed board hives in the form of a simple box, adding bars across the top so that the bees could attach the comb. This technique was never too successful because the bees usually attached their comb to the sides as well as to the top bar and the cover. The only way to obtain honey was to destroy the comb, often damaging the hive. Around 1806, a Ukrainian, Peter Prokopovich, a commercial beekeeper with about 1,000 colonies, produced what might be considered movable frames. The bars were notched so the bees could pass from one chamber to another. Honey was removed from the rear through a special opening. Even this hive was not satisfactory, as the bees still attached frames to the walls with wax. About the same time Greek beekeepers were experimenting with convex top bars and using the top of the hive to remove honey. Combs could be removed, but breakage was considerable.

Ship traffic was world-wide by the beginning of the 17th century, and, along with his horse, cow, and dog, man took his beehive. The honeybee was introduced to the North American continent about 1620. Twenty years later honey was quite abundant in Virginia. The Indians apparently had mixed feelings toward bees, referring to them as "white man's flies." One Indian was heard to say, "White man makes horse work, makes ox work, makes flies work; now time for injun to leave." Honey sold for 2 shillings a pound in 1650, and one report mentions an individual who owned 13 colonies. There is no evidence to indicate that the French brought bees to their colonies between 1699 and 1776.

Once established on the continent, the honeybee found the new environment quite suitable. Although no large fields of clover and alfalfa were available, there were trees, shrubs, and flowers that produced an abundant supply of nectar and pollen, such as maple, oak, willow, basswood, blackberry, cherry, sunflower, goldenrod, aster, and sumac, to name a few. A healthy colony, then as well as now, will swarm if it is not properly managed. Some 200 years before the use of movable frames it can be assumed that escaping swarms were quite common. The actual dispersal of honeybees in the United States, either by man or through swarming, is poorly documented.

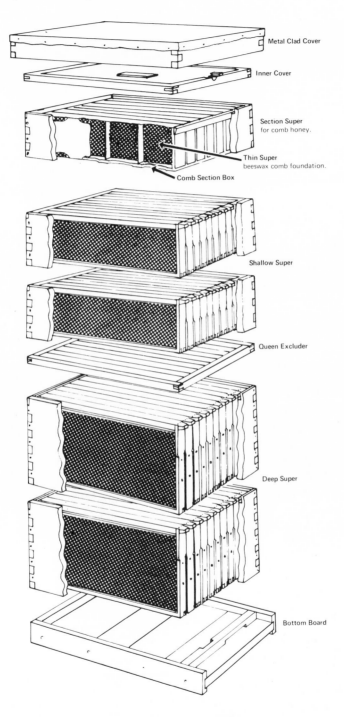

Metal Clad Cover

Inner Cover

Section Super
for comb honey.

Thin Super
beeswax comb foundation.

Comb Section Box

Shallow Super

Queen Excluder

Deep Super

Bottom Board

Components of a typical hive.

Bees could and probably did spread westward by four routes: through the Hudson and Mohawk River Valleys, along the Ohio River Valley, through the Cumberland Gap, and along the Gulf Coast. Each route had a supply of hollow trees and a succession of pollen and nectar plants that would have assured the wild colony adequate stores to survive the winter.

Honeybees arrived in Tennessee somewhere between 1748 and 1750; in Ohio, about 1754. Whether they arrived with man or by swarming is unknown. The 1850 census states that Tennessee produced about 1 million pounds of wax and honey. Kentucky also records a rather early arrival of the honeybee, 1780. Texas, Louisiana, Arkansas, and eastern Nebraska record the presence of honeybees between 1800 and 1820. At this time Indians sold beeswax in the vicinity of Chicago for 39 cents a pound. The 1850 census records bees and beeswax in 29 states as far west as Texas, Iowa, Wisconsin, and Illinois. Wax and tallow were listed on inventory records at Prairie du Chien, a small trading post at the junction of the Wisconsin and Mississippi Rivers, in 1818. Inventory records at Superior, Wisconsin recorded 10 pounds of wax valued at $2 about the same time. Indians, however, were probably collecting honey some time before these documented dates.

Indians in southern Wisconsin burned out trees and brush over relatively large areas to improve hunting and agriculture. A number of oak trees survived these prairie fires, and these areas are called oak openings. Goldenrod, asters, raspberries, blackberries, and basswood were undoubtedly plentiful; hollow trees were also abundant and provided natural beehives. It is believed that Indians collected honey before the arrival of the white settlers, evidenced by the fact that ladders were attached to a number of hollow trees. They were made from tall, thin trees whose branches were cut off, leaving prongs 8" to 10" long to serve as rungs. Early travelers described this part of Wisconsin as one large apiary. Hunting bee trees provided a profitable enterprise for the early settlers. There are many reports of trees yielding 25 to 50 gallons of honey. During this period honey sold for 25 to 37 cents per gallon. It was not unusual for a honey hunter to make $170 per season. That honey was indeed plentiful is attested to by the fact that there are several rivers named Honey Creek and Honey Lake. A Beetown township and a village of Beetown offer further proof of this unique heritage in southern Wisconsin.

Normal sealed brood and three swarm cup cells. When the colony becomes crowded, the queen lays an egg in each cup cell. Workers will raise a new queen in anticipation of swarming.

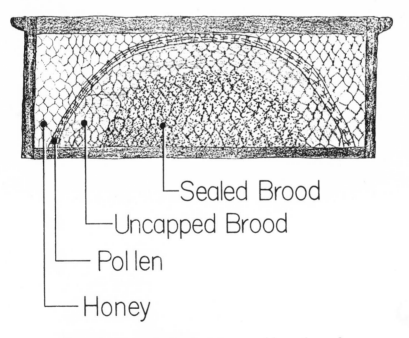

Sealed Brood
Uncapped Brood
Pollen
Honey

Organization of a typical frame of brood comb. The old brood is in the center, surrounded by uncapped larvae. Pollen is stored on the sides and top of the brood. Honey is stored on the sides and above pollen.

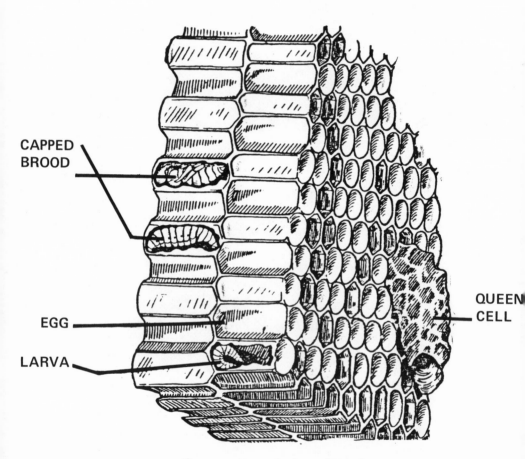

CAPPED
BROOD

EGG

LARVA

QUEEN
CELL

Cross section of a brood comb.

The Development of Modern Beekeeping

By 1851 progressive beekeepers knew enough about bees to manage them successfully, but they still had no truly suitable hive. A hobby beekeeper, Lorenzo Lorraine Langstroth, discovered and patented the idea of bee space in 1851. Rev. Langstroth developed an interest in insects as a youngster and later in life, when visiting a friend who kept bees and seeing some of them in a glass globe filled with honey and comb, his interest was aroused. He purchased two colonies with what was then conventional equipment and enlarged the grooves into which the top bars fit to allow 3/8" between the top bar and the underside of the cover. He noted that bees did not build comb in this space, so the cover could be removed easily. Langstroth was an alert observer and associated this 3/8" space with the lack of comb. Frames began to be designed to allow a 3/8" space on all four sides. The truly movable frame had been invented. Additional studies showed that if the space was less than 5/16", the bees would fill it in; if it was greater than 3/8", the bees would build brace comb. Equipment manufacturers now consider these facts in designing and fabricating equipment.

In the middle of the Civil War the United States Government made two landmark decisions that started an American Revolution in agriculture. In 1862 the Department of Agriculture was organized as an agency of the Federal Government. Part of its responsibility was to carry on research in order to solve agricultural problems. In the same year Congress passed the Morrill Act, which established the land grant state universities. These two events more than anything else affected the wellbeing of the average American citizen by providing him with a low-cost, abundant supply of healthful, nutritious, and varied food. These events affected beekeeping. In 1885 scientists in the Department of Agriculture began researching bee problems. The results have had a worldwide effect. Utilizing resources afforded them by the Morrill Act, some states began teaching beekeeping as an academic subject and began conducting research on specific local problems. These programs continue today. Today, there are six Department of Agriculture field-research laboratories involved with bees, some of which are operated in conjunction with state universities. Each laboratory has one or more primary missions, such as pollination, colony management, disease control, and breeding. They are staffed with trained professionals.

Bee diseases have been recognized for many years. Early efforts to control them involved mandatory inspection, quarantines, and in some cases burning the diseased bees and hives. The first bee-inspection law was enacted in 1877 in southern California (San Bernandino County). By 1883 the entire state had a similar law. Twelve states had enacted comparable laws by 1906, and currently most states have some regulations for bee diseases. At the Federal level the Bee Act of 1922 regulated the importation of bees from foreign countries. Controlled mating also interested researchers and queen producers for many years, but it was not until the 1920s that enough was known about bee anatomy and physiology to develop workable techniques for artificial or instrument insemination.

No summary of beekeeping history would be complete without mentioning the Dadant and Root families, both of which have made valuable and historic contributions to the beekeeping industry. Both families are well into their fourth generation and are still active in managing the business. Charles Dadant was born in France in 1817. He developed an interest in bees at about the age of 12. The family emigrated to Illinois in 1863 and took to beekeeping as a family occupation. Charles began importing queens from Italy in 1868, making several trips to arrange large-scale importations. He was convinced of the merits of the Langstroth hive and his writings in French and Italian introduced it to Europe, the home of the honeybee. In 1871 his son, Camille Pierre, took over the operation of the business and began making foundations for his own use and later for sale. The business prospered and expanded. In 1885 Langstroth asked the Dadants to revise his famous book, *The Hive and the Honeybee.* Charles translated this book into French, and later it was translated into Italian, Russian, Spanish, and Polish. The *American Bee Journal* has been published by Dadants since 1912. As the company prospered and grew, new inventions were added to the business. The fourth generation of the family took over a modern and complex business with many interests. The company has developed plastic-based foundations and hybrid queens. It is the world's largest manufacturer of beekeeping equipment and supplies. Foundation is the thin sheet of beeswax on which are embossed three-faced bases of honeycomb cells. It is held in the hive by frames. This increased business and the company prospered.

In August 1865 in Medina, Ohio A.I. Root, who was a jewelry manufacturer, spotted a swarm of bees overhead. A workman asked what he would give for them, and, to the workman's astonishment, Mr. Root offered him $1, probably more than a day's wages. To Mr. Root's amazement, he had just purchased a swarm, since the workman was able to capture and deliver the bees in a makeshift wooden box. Root began asking questions, reading, and studying bees. He recognized the value of the Langstroth hive and of movable frames and started his business exclusively with them. News of a honey ex-

Bees can be packaged and shipped long distances.

tractor came from Germany in 1867, so, using Langstroth hives and a mechanical extractor, Root extracted 1,000 pounds of honey from 20 colonies. His first year's winter losses were severe, but he was not discouraged, increasing his operation to 48 colonies the following year and producing 6,162 pounds of extracted honey. Honey at that time was 25 cents per pound.

Other people wanted to get into the beekeeping business, and much of the equipment available at that time did not suit Root. He began manufacturing his own and selling it to others. To answer questions asked by beginners, he sent out circulars free of charge. In 1873 he decided to issue a quarterly publication that sold for 25 cents per year. This became quite popular, so he expanded it to a monthly magazine with a subscription cost of 75 cents per year. He called this magazine *Gleanings in Bee Culture* — the same title that it has today. Root apparently lost interest in his jewelry business. By 1876 he had doubled his work force in the bee-supply business and had to run the plant day and night during the busy season to fill orders. By 1883 he had again doubled its capacity. Company expansion continued, and today production facilities cover 183,000 square feet. Other manufacturing facilities are now located in Texas and Iowa. Company contributions were significant. They publicized the point that it was possible to make a living by keeping bees. Root led the movement to standardize equipment as opposed to individual sizes and shapes of hives. The company's interest in helping others get started in the beekeeping business led to the development of a convenient container in which to ship bees. This innovation led to the bee-packaging industry. Marketing honey was another large part of the Root family business. By 1920 they were advertising nationally. This phase of the business was sold in the late 1920s, and the company, under the guidance of H. H. Root, A. I. Root's son, entered the church-candle business. The company invented and patented a rolled candle that does not bend in hot weather. Today the Root Company has supply stores in scattered locations throughout the United States and representatives selling church candles directly to users.

While history never ends, it is only fitting and proper to say that in the early 1970s there was a dramatic surge of interest in beekeeping. An individual need not be an apiculturist to make a living by keeping bees, nor even an agriculturist who needs bees to pollinate crops. You need only be interested in bees. You may be a professional or a businessman who wants a change of pace, or an individual who enjoys the challenge of working with insects. You may like honey, but your primary interest is probably the plain little honeybee, *Apis mellifera,* who first attracted man's attention in antiquity and still does today.

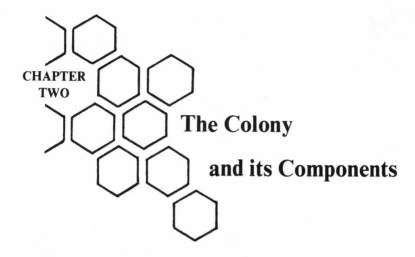

The Colony

and its Components

BEES ARE NOT THE ONLY social creatures — ants, termites, and a number of wasps are also social — but it is generally agreed that the honeybee is undoubtedly the most highly developed. There is a well-marked caste system, with the queen differing substantially from the workers. In less developed social groups the queen is able to carry out most functions to some degree, usually building the nest and gathering food, but when it is time to lay eggs, workers attend to her needs. The honeybee queen is quite helpless or so specialized that she cannot exist by herself, but, she is an extremely efficient egg-laying machine, capable of 1,500 to 2,000 eggs per day.

The colony has three component parts or individual types — workers, queen, and drones. The life span of the colony is perennial, but that of the individual varies. The queen normally lives 2 to 3 years (up to 8 years according to some records), workers 4 to 5 weeks, and drones about 30 days.

Workers

The workers are the smallest members of the colony and comprise the bulk of the population: a full-size colony may have up to 60,000 workers in peak season, although during late winter the number is considerably smaller. Workers are undeveloped females: each worker originally has the potential of becoming a queen. Her ovaries are undeveloped, but she has other organs and structures necessary to carry out her many duties for the well-being of the colony.

Workers develop from fertilized eggs. The queen usually deposits one egg in each cell. It is glued to the vertical surface on one end. After 3 days the larva emerges and is immediately flooded with royal jelly, which is secreted by worker bees. After 2½ days the diet is changed for the next 2½ days to include some pollen and honey. A newly hatched larva floats in its food. As it grows, it remains curled in the cell. After it is fully grown, the cell is sealed with a porous wax covering. The larva then stretches out, head pointing outward, spins a cocoon, and pupates. It now goes into a quiescent stage, while vital changes occur internally. Eyes, legs, wings, and a number of organs gradually take shape. After three weeks the adult worker chews her way out of the cell and is ready to go to work.

The length of the individual worker's life varies considerably, depending on the time of year. Workers emerging in the fall can expect to live the longest. Those emerging in early spring live on the average about 35 days, while those who emerge in or just before the honey flow have the shortest life expectancy — about 28 days. Such differences are undoubtedly associated with the wear-and-tear on their bodies. The hardest work in terms of shortening life expectancy is brood rearing — i.e., feeding larvae. The worker occasionally depletes protein and fat reserves in her own body, especially if pollen is not available in sufficient quantity. Brood rearing normally slows down in the fall. Workers in the hive at this time tend to accumulate fat and protein in their bodies and therefore live longer, which helps assure colony survival through the winter. While we may view foraging for pollen and nectar as a hazardous occupation (which it undoubtedly is), bees that forage early in their life have a shorter life expectancy by 7 days compared to those that began foraging later in life.

Many factors affect the behavior and activities of honeybees. Age is one of the most important: a worker cannot secrete wax or build combs until her wax glands are mature. She cannot sting until that mechanism is fully developed, and venom secreted. Some behavior is genetically controlled: some workers uncap only cells that contain dead brood; others with a different set of genes do not uncap cells but do remove dead larvae. It is obviously desirable to have both types in a colony. External stimuli also influence honeybee behavior. The body is covered with hundreds of sensory hairs and receptors, all of which are connected with nerve cells, many with fixed pathways. A specific stimulus brings about a specific and often very rapid response. Internal control mechanisms are also present. Many animals and insects

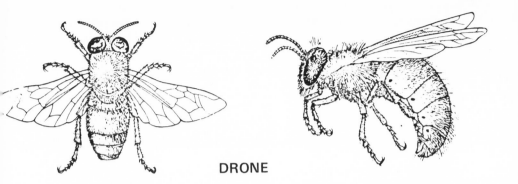

DRONE

Differences in size and profile of three members of the honeybee colony.

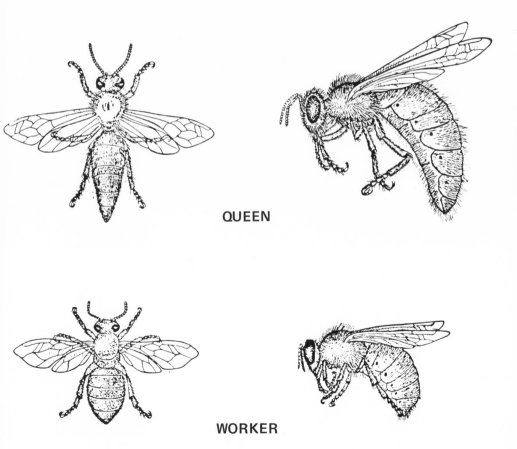

QUEEN

WORKER

have a so-called built-in time clock, on about a 24-hour cycle. Bees tend to forage during a specific time of the day, often synchronized with nectar or pollen secretion. Bees as well as other animals and insects have patterns of behavior. For example, if the pollen pellets that a foraging bee is carrying into the hive are removed at the entrance, she will continue to go through the routine or ritual of searching out a pollen cell and unloading the pellets from her legs.

Many people wonder how bees know what is needed in the colony. Their activities should be interpreted in terms of the known facts. To follow the path of a worker bee stimulated for some reason to secrete wax: as she moves about the hive, her sensitive legs or antennae can detect a cell that needs repair or one that should be capped. She will respond or react to this stimulus to a greater degree than would a bee just returned from a foraging trip. The same principle is probably true for workers feeding larvae, cleaning cells, or performing other tasks. To the casual observer such behavior might appear to be under the direction of some administrator or supervisor. Honeybees have a memory and are capable of learning. Those routinely foraging on certain crops learn the most efficient way to seek out nectar. If a hive is moved from one location to another, the field bees invariably return to the old location and look in vain for their home. Beekeepers use specific tricks to move hives short distances.

Much has been learned about the behavior of honeybees by marking or numbering them and watching their activities through observation hives. For the first 3 days after emergence workers usually clean cells from which other bees have emerged and feed older larvae. After workers reach 6 to 12 days of age, they begin to feed young larvae. During the third week worker activities become more varied. They may take an orientation flight, sometimes referred to as play flight. During these flights they apparently learn to recognize their own hive and to locate it in relation to nearby surroundings. Once the workers reach 3 weeks of age, they are capable of performing many different kinds of services, depending on what needs to be done. Some would say that they respond to stimuli that suggest what the colony needs. Bees of all ages respond to or are involved in temperature regulation. The brood nest is maintained at a temperature of 95° F. (34-35°C.). They cool the temperature by fanning and water evaporation; in the winter heat is retained by clustering and generated by utilizing honey as energy.

Specific activities that honeybees engage in may be changed quite frequently, depending on the needs of the colony. The term "busy as a bee" was apparently derived from philosophy, not science. Dr. M.

Lindauer observed a number of marked bees over long periods of time — one that he watched for 68 hours and 53 minutes was inactive for the entire period; another worker spent 56 hours, 10 minutes examining cells but not apparently doing much work. With the exception of foraging the same activities appear to be performed day and night. Do bees sleep at night? Apparently not, but they do rest at randomly distributed time intervals.

Secreting Wax and Building Combs

Wax is secreted by glands on the abdomen. They are most productive when bees are between 12 and 18 days old. Wax is secreted in the form of flakes, which project from the overlapped portions of the last four abdominal segments. Considerable honey consumption and relatively warm temperatures are required: secreting wax requires considerable energy. About 8 pounds of honey and some pollen is needed to produce 1 pound of wax. Workers actively secreting or about to secrete wax gorge on honey and often hang in groups in the vicinity of comb building. It may take up to 24 hours from consumption to actual secretion.

Wax is removed from the abdomen by one of the hind legs and passed to the mouth, where it is chewed. Secretions from the mouth are undoubtedly added, because wax flakes that drop off to the bottom board look different than comb. The entire process of removing, chewing, and affixing one scale takes about 4 minutes. Bees are not the most efficient workers: one bee will affix a scale; a moment later another will come along, gnaw it off, and move it over. Particularly during capping, workers do not hesitate to borrow a bit of wax from a nearby cell to finish the job. They are apparently very sensitive to gravity and always construct the comb vertically. Once movable frames were discovered, Johannes Mehring invented wax foundation, which helped assure straight combs.

Attempts were made to fabricate foundation and comb from material other than beeswax: aluminum foundation coated with beeswax has been used; completely plastic combs (some coated with beeswax) are available. Bees will use them, but if they are given a choice (such as using half plastic frames and half natural), they prefer the natural. A good synthetic comb would have distinct advantages in commercial honey production: honey could be extracted more rapidly by mechanical means, and combs infected with a disease could be heat-sterilized quite easily.

Feeding and Brood Rearing

Workers can begin feeding newly hatched larvae when they are about 3 days old and continue for about 12 or 13 days. The food that the larvae receives is a secretion produced by the brood food glands called royal jelly. The same mixture is given to the queen. In order to secrete this highly nutritious food, the worker must consume pollen. If pollen is temporarily lacking, she will draw on body reserves. Once these body reserves are gone, brood rearing stops unless additional pollen becomes available.

When workers are about 12 or 13 days old, they stop feeding larvae. They can continue nursing in emergencies, but they usually find something else to do. The feeding visit may be very short (2 or 3 seconds) or take longer periods of time. Observations by Lindauer, using marked bees, noted that 2,785 bees spent 10 hours, 16 minutes, and 10 seconds taking care of the cell and larva until it emerged. Studies using radioisotope-marked and colored food indicate that food is liberally passed from bee to bee in the hive. Feeding is generally very rapid. During feeding antennal contact is made, and possibly other forms of communication are transmitted. Other chemicals such as odors are undoubtedly also exchanged.

Protection

Everyone knows that bees sting, but colony defense is not completely understood even by some professional beekeepers. During the honey flow fewer bees guard the entrance than after the flow is over. Workers loaded with pollen or nectar are freely admitted. Guard bees usually stand at the entrance on four legs; the two front legs and antennae appear poised to challenge intruders or strangers. Older bees entering their colony appear to ignore the guards, but younger ones often submit to examination or inspection. Stinging is the defensive behavior exhibited to intruders. It is the worker's reaction to specific stimuli. For example, bees will strike at a rapidly moving object but will ignore this same object if it moves slowly over the same area. Bees will strike at a black or red object before a white one; at a shiny object such as a wrist watch or glasses rather than bare skin. More bees will take up a defensive position on a cloudy day than when it is clear and bright. Bees are more defensive and more likely to strike when no pollen or nectar is coming in than during the honey flow.

Once a bee strikes or stings, an alarm pheromone (isopentylacetate) is released. A pheromone is a chemical released outside the

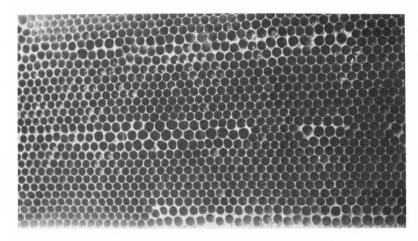

Drone cells. The queen will lay an unfertilized egg in each large irregular cell, and a drone will develop.

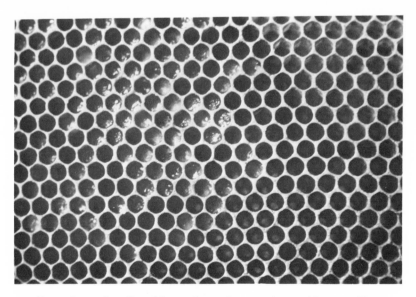

Brood comb. Good brood comb contains uniform cells. Pollen is stored for future use.

body to which other individuals of the same species react. There are sex pheromones as well as other types. If a bee is accidentally crushed while manipulating frames, several bees will often react violently and attempt to sting.

Smoke has been used for generations to quiet bees. Beginning beekeepers learn very quickly the value of smoke in handling bees, especially if they have tried to manipulate or examine a colony without it. Smoke stimulates bees to begin filling up with honey, and for years it was believed that this was why colonies could be easily manipulated. It is now thought that smoke in addition to stimulating engorging masks the odor of the alarm pheromone or dulls the senses of bees so that they do not respond to it.

The stinger of the honeybee, as well as those of wasps, hornets, and yellowjackets, is a modified ovipositor (i.e., a structure designed to insert eggs into specific areas). The stingers of wasps, yellowjackets, hornets, and bumblebees are smooth, so they can sting an individual several times in quick succession. The stinger of the honeybee is barbed and cannot be retracted from bird or mammalian skin without fatal injury. She can sting another insect or bee and not be injured. Beekeepers wearing protective clothing often receive a number of strikes in which the stinger is torn from the worker's body, yet the injury is not immediately fatal to the bee. She will continue to buzz around the person, creating the impression that many bees are vigorously defending the colony when in fact only a few may be doing so.

All stings are painful and quite uncomfortable. To a few individuals (less than 1%) they can be dangerous and even cause death if proper medical help is not administered quickly. Public-health records indicate that stinging insects were responsible for 229 deaths between 1950 and 1959, more than rattlesnakes. Wasps and hornets were responsible for 101 of these, bees 124, and ants 4. These statistics do not take into account the amount of exposure: more people are exposed to bees than to all other causes of death combined. Millions of people are probably stung by bees and wasps each year, and thousands experience a painful reaction, but records and reports of such incidents are lacking. In most situations people fail to or cannot distinguish between wasps, hornets, yellowjackets, and honeybees. Is it possible that some deaths are officially recorded as heart attack or heat stroke when they may have been due to a sting? Apprehension is caused if a stinging insect enters a moving automobile. Have any fatal accidents resulted when a bee, wasp, or hornet entered an automobile and caused a momentary distraction?

The medical profession categorizes reactions to insect stings into three groups: (1.) Hymenopterism vulgaris. This is the most common reaction, experienced by the majority of people stung. An immediate intense pain is followed by localized swelling and reddening of the area. Swelling can impair the normal functioning of a joint, so one might have a stiff finger. Symptoms may last from an hour to several days. (2.) Hymenopterism intermedia. This is an arbitrary classification in which the arm or leg could swell, for example. Stings on the lip, nose, or tongue are put in this category. (3.) Hymenopterism ultima. This is a near-fatal or fatal reaction if medical help is not received in a relatively short time. Physicians recognize this as anaphylactic shock. The individual may break out with hives, with shallow breathing and pulse and heartbeat greatly reduced.

Venom from stinging insects is a complex mixture. Formic acid accounts for the immediate intense pain; other ingredients are responsible for the drop in blood pressure and respiratory failure. One ingredient, phospholipas A (an extremely toxic component), is also present in Indian cobra venom, known as a neurotoxin. Bee venom contains melittin and apamin, which increase plasma cortisol. This might explain the old wives' tale that bee stings help alleviate the symptoms of some arthritic-type conditions.

People vary in their sensitivity to bee stings. Families of beekeepers (wives and children) have a higher than normal percentage of sensitive individuals. Perhaps they became overly sensitized by the low-level exposure. The potency of a bee sting varies with the season and even with the plants that the bees are visiting, although there are differences of opinion here. The quantity of venom injected is also variable. You might receive the whole load or only a small percentage. Sensitive individuals can be immunized or desensitized. Such procedures should be done only under competent medical authority.

In addition to defending herself against intruders, the worker is also able to defend herself against inclement weather — too hot or too cold. Insects are normally thought of as cold-blooded creatures, since their body temperature is near that of the surrounding environment. The honeybee differs. It resembles warm-blooded creatures in that its development is predictable: i.e., 21 days from egg to emergence. If the cluster is considered as an organism rather than an individual bee, the colony is a warm-blooded creature, because it is maintained at a relatively constant temperature of about 95°F. (34-35°C.). If the surrounding environment is too warm, workers will begin fanning air. Air currents help evaporate moisture from honey and also regulate the temperature of the hive. The honeybee is able to survive temperate

BEE DANCES

Wagtail
Food source directly toward sun.

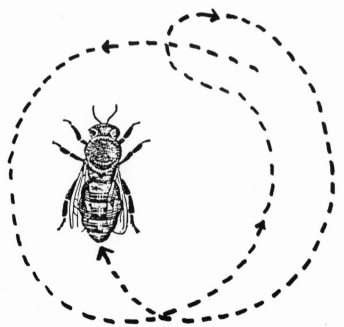

Round
Food less than 300 feet from hive, no direction given.

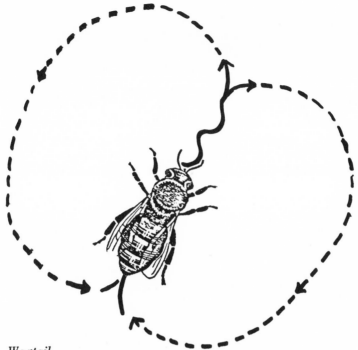

Wagtail
Food source to right of and toward sun.

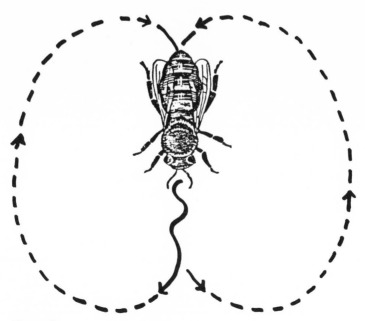

Round
Food source in opposite direction from sun.

climates because it can regulate temperature. Studies have shown that the temperature inside the winter cluster remains fairly constant. Bees on the outside act as insulators, keeping heat inside. Heat is generated by metabolizing honey. Placing hives in artificially low temperatures has no detrimental effect as long as honey is available inside the cluster. As the temperature rises, the cluster expands and moves to new supplies. Once the temperature is as high as 57°F. (14°C.), foraging bees will venture away from the cluster and start moving honey around the winter cluster.

Foraging

Bees venture out and start flying when they are 3 or 4 days old, but field activity begins fully at about 3 weeks. They collect nectar, pollen, water, and propolis, a resin or gum secreted by certain plants. This material is used to seal cracks and crevices in the hive. Sealing cracks in the hive actually glues together hive bodies. Beekeepers do not like colonies that gather large quantities of propolis, as it is difficult to remove and time-consuming. Water is collected to dilute honey fed to brood and to cool the hive. Beekeepers occasionally feed granular sugar to bees, and water is also collected for this. Water and propolis are collected as needed and are not stored for future use.

Nectar is the basic ingredient of honey. It is a sweet material, composed primarily of sugars and produced or secreted by flowering plants. Some call it an excretion; others label it as the necessary attractant for pollination, which assures propagation of the species. Nectar is the source of carbohydrates for bees. When it enters the honey stomach (crop) of the bee, it is mixed with an enzyme, invertase. As mentioned previously, bees exchange food quite liberally in the colony. When a field bee loaded with nectar enters the hive, she is greeted by guards and other workers. She will regurgitate nectar to a number of workers until her crop is empty, then return for another load. In most cases she will return to the same vicinity, often to the same plant species. Meanwhile, the nectar has probably changed crops several times, each time being mixed with more invertase. When hive workers have received more nectar than they need, they will deposit some in an empty cell to the side or on top of the brood nest. Other workers may decide that the queen needs the nectar for an egg. They will ingest it and move it to another cell higher in the hive. Whenever a transfer or move takes place, an additional amount of invertase is added and moisture normally is reduced. When workers walking over the cell detect or receive the stimulus that the nectar is somewhere in the vicinity of 20% moisture, they cap it with wax. Nectar starts out primarily as a double sugar (12 carbon atoms); it is split into two simple sugars, glucose and fructose, by the enzyme invertase. After this happens and water content is reduced to about 20% or less, the product is called honey.

Some insects (aphids and scale) excrete a material high in sugars. Bees collect and store it as readily as nectar. Even though it is recognized as an insect excretion, the dew honey that results is wholesome. These aphids and scale insects ingest large quantities of plant sap which contains more sugar than protein, to meet their nutritional requirements. The excess sugars are excreted, and opportunistic bees collect the material for their own use. In some parts of the world honeydew comprises a large percentage of the honey crop. As a rule it is darker and has a distinctive and sometimes characteristic flavor of a specific crop or plant. It contains more gums, dextrins, and probably salts. Some beekeepers maintain that it is not good overwintering food for bees because of these extraneous, indigestible materials and salts.

Pollen is collected by honeybees and used as a source of proteins, fats, vitamins, and minerals. While she is in the process of collecting pollen and nectar, the bee accidentally carries pollen from one plant to another. It is not because of her good nature that crops are pollinated, but because the honeybee over millions of years became a vegetarian. Her probable ancestors (wasps and hornets) are carnivorous and/or scavengers. Bees collect some obviously indigestible materials such as road dust, sawdust, or cracked corn from bird feeders when the supply of pollen in the hive is low and none is being produced by plants. The urge, instinct, or stimulus to supply the required protein for the colony is undoubtedly very strong.

Pollen is produced by plants during specific times of the day. Pollen-gathering bees adjust their foraging activity to synchronize with pollen production. Younger bees generally gather pollen and usually work one plant species at a time, though exceptions do occur. The worker moves over the plant and often appears to be wallowing in the pollen. She then hovers over the flower, combing or brushing her body with the combs on her legs and packing the pollen in the pollen basket on her hind legs. Saliva, possibly nectar, is mixed with it in order to pack it better. Upon arrival at the hive she enters the brood nest, seeks out a suitable cell, backs in, and dislodges the pellets. Using their heads as battering rams, other bees now pack the pollen tightly into the cell. Honey cells are filled to the top with nectar; pollen cells, only about three-fourths full. Once pollen is packed into the cell, it undergoes a type of fermentation and keeps without spoiling. Man now utilizes the same type of device and technique to preserve forage for cattle in a silo.

As a general rule bees forage in a somewhat restricted area of a field and on a specific plant species. Less than 3% of bees carry mixed pollen loads back to the hive. As pollen and nectar become scarce, the foraging range widens. The number of plants visited varies from 8 to over 100. The bee may need anywhere from 8 to 187 minutes to obtain a full load and may spend from 4 to 150 minutes to unload, depending on circumstances in the hive. She may make as many as 24 trips in a day; the average appears to be 10 trips. Load weights vary: the average nectar load weighs about 40 mg. (1 milligram is 1/1000 gram, and 453 grams equal 1 pound), but some may carry up to 70 mg., which is 85% of the bee's weight. Pollen loads vary in weight and bulk density: it may weigh anywhere between 8 and 29 mg., depending on the plant species. The lowest reported temperature at which bees venture out for nectar is 55°F. (13°C.); 47°F (8°C.), for pollen. Foraging for pollen drastically diminishes at temperatures above 95°F. (35°C.), for nectar, above 110°F. (43°C.). At these temperatures they will still go out for water. Bees appear to stop gathering nectar at winds of about 15 m.p.h.; pollen, about 20 m.p.h. Wind velocity is difficult to measure, as there would be a considerable difference near the ground in the immediate vicinity of the plant. Loaded bees return to the hive at an average speed of about 15 m.p.h., which can range between 13 and 18 m.p.h. The return-trip speed averages 1 to 2 m.p.h. less.

Communications

The basic requirement for a social existence is some effective means of communication. Without communication there would be no way to transfer information, and this information should be capable of inducing a response in another receptive individual. The stimulus can be transmitted by light, chemical, physical, or possibly electrical means. Man is highly dependent on light or visual stimuli; since honeybees naturally live in a closed, dark environment, it is obvious that they depend on odor, touch, and sound stimuli to a greater extent than on light.

In 1788 Spitzner first reported that bees communicate information by performing the bee dance. His work was largely overlooked until K. von Frisch developed an interest in bee communications in 1920. He was awarded a Nobel Prize in 1973 for his spectacular achievement. Von Frisch established feeding stations around the

A swarm unable to find a suitable cavity will build comb in a sheltered area outdoors. Such colonies usually die during cold weather.

Queen cells. If the queen is removed, dead, or killed, workers select an egg or a very young larva from the regular brood area and raise another queen.

vicinity of an observation hive. By moving the station he was able to break the code, or translate bee language into people language. He described two types of dances, the round and the wagtail dance. These dances are performed in total darkness, so stimuli other than visual are being transmitted.

Communicating direction is of vital importance to the successful survival of the colony. Gravity and the position of the sun are the reference points. If food is found toward the sun, the wag-tail dance is performed upward on the comb; if away from the sun, the dance is performed downward. Intermediate locations are indicated by angular deviation from the vertical. The returning forager bee makes corrections for wind and terrain. As far as is known, bees cannot communicate up and down directions: for example, they could not inform other bees that a source of pollen was located on the eighth floor of a building.

Accuracy of communications was studied by several individuals. In one test of 150 marked bees that followed a dancing bee, 91 brought back nectar or pollen; only 79, the same kind of nectar or pollen that the dancing bee had previously collected. Of these bees, 42 now performed the dance, giving the correct direction and distance. Some may consider this a poor record, but so far as survival of the species is concerned, how desirable would it be for the colony to send all available foragers who received the message to the same spot for more? Is it not more advantageous for some of the recruits to locate other sources of food?

The round dance is performed when food sources (nectar and/or pollen) are within 100 m (about 300'). No direction is given. In the round dance the bee runs, using short steps, in a small circle. She changes direction so that she rushes once to the right, once to the left, and once or twice again in either direction. She may continue to dance for several seconds up to 1 minute at one location. She may then move to another part of the comb for a repeat performance, then rapidly leave the hive for another load. If food is further than 100 m away, the wagtail dance is performed. The foraging bee makes a half-circle, turns sharply, runs in a straight line to the starting point, and then makes another half-circle in the opposite direction, thus completing a full circle. She again runs in a straight line, retracing the same straight-line path and repeating the half-circles. While traveling along the straight line, she wiggles her abdomen in a sidewise motion, hence the term "wagtail" dance. During the straight-line movement she undoubtedly transmits distance information. One can view it in daylight and count the number of wiggles per unit of time or distance, but it should be kept in mind that this dance is performed in the dark: the bee could be transmitting sound or electrical impulses to other workers.

Behavior of the Colony as a Unit

A swarm rather than an individual bee probably first caught man's attention. At first glance the swarm might appear to be a rather haphazard accumulation of bees, but it is now known that there is order and harmony within the swarm and/or colony. In studying and describing insect behavior, one tends to interpret it in terms of human values and characteristics. Bee senses are interpreted in terms of vision, taste, smell, and hearing. It is known that bees respond to gravity and possibly to electromagnetic stimuli. Could there be other stimuli to which they respond? It is not out of the realm of possibility. Some writers describe the activities or work that bees do in the colony as a division of labor. In one sense this is true, but the idea implies a hierarchy — a dictator or at least a number of administrators — to make the system work. Known facts indicate that this is not true.

Swarming has already been mentioned as a means by which a colony reproduces. A definite sequence of events precedes swarming. Many a beginning beekeeper has observed the sequence without knowing what was going to happen, then much to his surprise found a swarm nearby. Colonies vary in their instinct to swarm. Some believe this to be genetically determined. Once swarm cells contain a developing queen, the field workers slow down and sometimes stop foraging. If the hive is inspected, one can see that it is extremely crowded with gentle or very docile bees interested primarily in resting in all available spaces. The queen has stopped laying. The bees are fully engorged with honey in preparation for an anticipated flight. On a warm day a large number (no fixed percentage) of workers will rapidly leave the colony. They will circle in a nearby vicinity until the queen arrives. When she is located, they emit a pheromone that attracts others to the location. The cluster will now move off to the new home. If no suitable domicile is found, the swarm will alight on any convenient structure, such as the limb of a tree or a fence post, and await a message from scouts. Swarms have been known to wait for several days before a suitable location is found. Most swarming bees are docile because they have no home to defend, but there are exceptions. A swarm unable to locate a suitable hive eventually begins to run out of food, and some workers will defend the swarm and sting.

Table 1. Mineral constituents of honey (in parts per million).

	floral	honeydew
potassium	205	1676
chlorine	52	113
sulfur	58	100
calcium	49	51
sodium	18	76
phosphorous	35	47
magnesium	19	35
silica	22	36
iron	2.4	9.4
manganese	0.30	4.09
copper	0.29	0.56

Table 2. Development and activities of honeybees (in days).

	queen	worker	drone
egg laid	0	0	0
egg hatched	3	3	3
cell capped	8	8	10
emergence	16	21	24
first flight*	18 - 25	31 - 40	30 - 35
life expectancy*	2 - 7 years	45 - 180	30 - 40

*considerable variation depending on conditions in colony

A worker bee can live on her own for 1 or 2 weeks, but not willingly. Stray bees are attracted to other groups or join another colony located by sight or scent. A large group of workers can form a group or cluster even without a queen, and some individuals will begin specific activities, such as foraging for nectar or pollen and comb building. The presence of a queen is important in maintaining colony organization, cohesion, and high morale, which is distinctly separate from the essential function of reproduction.

It was recently shown that secretions from a specific gland in the queen's mandible attract workers. The material was analyzed and found to contain fatty acids, 9-oxodecenoic acid, and 9-hydroxydecenoic acid. These chemicals are now referred to as queen substance. Queen substance prevents worker bees from attempting to rear new queens and also has a suppressing effect upon ovary development in workers. In the absence of a queen or the queen substance, workers immediately begin to build queen cells and to raise a new queen. In some cases the ovaries of workers begin to develop, and they may even lay a few eggs. Since these eggs are not fertilized, the resulting offspring are drones. The absence of a queen or the queen substance is probably noticed within 1 hour, and workers begin constructing queen cells within 5 or 6 hours.

Drones

In higher animals fertilization is generally required before an egg begins to develop. Several groups of insects (aphids, lice) do not require fertilization, and all individuals become females. With the honeybee the situation is different. Fertilized eggs develop into workers or queens, while unfertilized eggs become drones. Drones complete development in 24 days. They have half the number of chromosomes that females have: a drone has a grandfather and a mother but no father. Drones are larger than workers but shorter than a laying queen. Since they are males, they have no stinger, which is a modified egg-laying structure. The tongue is short and they do not collect nectar from flowers. Scent and wax glands are missing, but odor-detecting organs are highly developed. The drone is a highly specialized creature; his only duty is to mate with a queen when one becomes available.

The normal colony begins to rear drones in late spring and early summer. The number produced depends on the size and condition of the colony. A small or weak colony produces fewer and sometimes no drones. The condition of the comb influences the number of drones: damaged comb is sometimes repaired or reconstructed into drone comb. When food is plentiful, workers feed and tolerate drones, but as soon as it becomes scarce, workers in a normal colony stop feeding drones. As a result many starve; others are slowly pushed out to die.

The Queen

For centuries naturalists observed bees, but not until the 17th century was the queen's true role discovered. Some early observers recognized her as the ruler, because if she were missing, things did not run smoothly in the colony. In 1609 Reverend Charles Butler announced that the king bee was a female and should be referred to as the queen. Not until the mid-18th century was the true relationship between drone and queen understood.

The queen is readily distinguished from other members of the colony. She is larger than workers and longer than drones. Her wings are shorter in proportion to body length than those of either worker or drone, but they are in fact, larger than those of workers. The abdomen is more pointed than that of a worker, and at a quick glance she looks more like a wasp. Movements of a mature, laying queen are slow and deliberate. A laying queen is generally in the vicinity of young brood. She is surrounded by a court of six to ten workers, who examine her with their antennae and continuously feed and groom her.

If the queen is suddenly removed or killed, workers will immediately begin raising a new one. No one knows exactly why a specific egg is selected. Upon hatching, the larva is fed royal jelly for 5 days. During this period the developing larva grows larger than a normal worker, so the cell is modified. The force of gravity apparently bends the cell downward, so the larva develops from a large cell that points downward rather than horizontally, as those of workers and drones. There are differences of opinion as to whether the queen results from an overabundance of protein-enriched food or whether workers result from deliberate underfeeding. At any rate the basic difference between a queen and a worker is nutritional.

It has been shown experimentally that intercastes can be developed. An intercaste is neither a true queen nor a true worker. If larvae over 3 days old are selected to become queens, imperfect queens usually result. The best queens are selected before they are 2 days old and raised in a strong colony. A new queen is also produced when conditions in the hive become crowded. The exact mechanism is not fully understood, but the queen will deposit an egg in the swarm cell. Workers feed these newly developing larvae a special diet for 5 days. The cells are sealed when the larvae are mature. Development is complete in 14 days. Developing queens are not necessarily the same age, nor, for some unknown reason, does the first queen that emerges survive, but hopefully the best one does. Whether the queen is reared in anticipation of swarming or because the old queen was killed or removed, the results are genetically identical.

After the queen emerges from the cell, she seeks out other queen cells. She makes an opening, inserts her abdomen, and stings her rivals to death. Workers dispose of the carcass and often tear down the remains of the queen cell. If swarming is anticipated by the colony, workers will gather around the extra queen cells and prevent the first queen from destroying them. This allows several swarms to emerge from one colony. From the time of emergence to her first flight the newly emerged queen is quite active. At first she is ignored by workers; later they appear to groom and even to attack her. She takes this mauling without resistance, occasionally emitting a shrill sound referred to as piping. Other queens still in the cell sometimes respond with what appears to be a different sound, but it may be muffled by the cell. Attacks stop after the piping. They appear to be made for conditioning purposes, since the queen is about to make a very important flight.

The first mating flight usually takes place 5 or 6 days after emergence. The new queen rushes excitedly toward the entrance. At first the guards ignore her, but later they push her out with their heads if she is reluctant to go. She will take several mating flights. After each, attacks by workers diminish, and, once egg laying begins, the workers stop attacking her. Observations of queen behavior suggest that there is considerable variation in the worker-new queen relationship. Nevertheless the entire colony is involved in producing a new productive queen. Mating flights take place between 1 and 5 P.M., with most occurring between 2 and 4 P.M. The better the weather, the greater the number of drones and the better the chances of a successful mating. Mating occurs at temperatures above 68° F. (20°C.). Queens that mate in unfavorable weather conditions receive little sperm and will usually be replaced. Egg laying can begin as soon as 14 hours after mating, but usually it occurs in 2 or 3 days. In 1 year the queen is capable of laying 200,000 eggs.

The queen bee is a unique individual not only because she can lay large numbers of eggs but also because she can control the sex of each individual. Before depositing an egg in a cell, she examines it with her head, antennae, and front legs. Some observers believe that her front legs serve as a type of caliper for determining whether a drone or a worker egg is to be deposited. The queen then backs into the cell and deposits an egg (covered with a mucilaginous glue) to the vertical wall of the cell. If it is a large drone cell, she lays an unfertilized egg; if it is a regular cell, the egg is fertilized. Under normal conditions she has absolute control and makes no mistakes. The queen normally slows her egg-laying rate in the fall and, by November or late December in the northern hemisphere, practically stops. It is generally believed that the phenomenon is regulated by the length of day (photoperiod). As the days begin to lengthen in January, egg laying begins again. The rate of egg laying is dependent upon the number of bees available to tend brood and keep the cluster warm and on the availability of pollen. If pollen is in short supply or not available to the winter cluster, egg laying will be slow or stop completely.

CHAPTER THREE

The Essentials of Beekeeping

BEES ARE SO INTERESTING and intriguing that you may wish to try your hand at keeping a few colonies. It is an interesting and challenging diversion, regardless of your occupation or profession. Some major discoveries about bees and beekeeping were made by individuals who kept bees not to earn their livelihood but as an avocation: for example, Langstroth was a minister; A. I. Root, a jewelry manufacturer; Charles Butler, another minister. This chapter is not a how-to guide for the beginner, but it does explain what is involved in becoming a beekeeper, or apiculturist. Beginners should read literature prepared by equipment suppliers, the U. S. Department of Agriculture, and state land grant universities. Most beekeepers readily share their knowledge and experience. You can easily meet other people interested in bees by joining local or state associations. This chapter reviews, recaps, and summarizes beekeeping procedures in the light of recent findings in apiculture. A more precise understanding of the whys and wherefores will enrich your experience and make beekeeping more exciting and pleasurable.

Where can bees be kept?

Bees can be kept practically anywhere. The basic requirements are a suitable place to locate the hives and a supply of flowering plants. Other than in downtown metropolitan areas of very large cities bees can and do survive and in fact can do quite well in smaller cities and suburbs. Several points should be considered. Keep the hives out of sight of the general public. Just the sight of beehives may frighten some individuals. The flight pattern of bees to and from the hive is also important. It should not be over someone else's outdoor clothesline, a busy sidewalk, a swimming pool, or a children's play yard. Flight patterns can be diverted or elevated to a degree with hedges of evergreen shrubs, walls, and fences. Bees can be forced to go up and over such barriers. At a higher elevation they are less likely to create a problem. It is advisable to have a fenced yard so that small children do not accidentally wander in front of the hive.

Keeping a colony or two in a metropolitan or suburban area requires more time and closer attention to details. Never work the bees on cloudy days, as they are more aggressive in defending their colony. They might fly a great distance from the hive and sting a neighbor. Be certain that the hive does not swarm — this could create panic. Provide water for the hives. A bird bath works well and reduces the likelihood of bees collecting water from your neighbors' swimming pool, bird bath, or even a nearby public drinking fountain. The rapport you establish with your neighbors determines the feasibility of keeping a colony of bees. Most professional beekeepers agree that bees can be kept in a metropolitan or suburban area. They also agree that they can be improperly managed and create a nuisance. Most cities have nuisance ordinances, and public health officials have occasionally declared beehives a public nuisance and ordered them moved or destroyed.

Bees routinely forage within a radius of 3 to 5 miles from the hive. In most areas within this radius there are lawns with clover, gardens, and sometimes parks containing flowering plants. In some suburban areas the nectar flow is better than in rural areas, especially in times of drought. Lawns and gardens are watered often, so plants are less likely to be dry.

Locating hives in a rural area or out in the country poses no special problems. A few items should be considered. Place the hives in a sheltered area. In northern climates hives should be sheltered from cold north winds if possible. In warmer areas it is desirable to place them so that they have afternoon shade. It is also more comfortable for the beekeeper working the hives. As a general rule bees in hives located in continuous shade tend to defend their colony more aggressively than in those exposed to sun. If possible, have the hive entrance face south or southeast. Bees in these colonies will begin work earlier in the morning. In winter they can take a cleansing flight more often than if the hive faced north. If you live on a farm, the south side of a building is an ideal place to locate the hive. If your hives are located away from the farm, it is highly desirable to be on or very near an all-weather road. Many colonies are lost in late winter or very early spring because the beekeeper could not get to his hives.

How necessary is it to locate the hives next to the crop from which you want to obtain honey? Bees are opportunists: they will work the crop that suits them best. They have been known to fly over an apparently good source of nectar to reach something more desirable. While bees do utilize energy in flight, whether they fly ¼ or 1 mile is not too significant.

How to begin Beekeeping

The first basic rule is to start with at least two colonies. The rationale for this is quite simple: if one of your queens dies, you can always combine the two colonies. The probability of both queens dying is quite remote. Although the bees will raise another queen, they can do so only if eggs or very young larvae are present. In northern areas temperature is another factor. Even if a new queen is produced, she will not venture out and mate unless the temperature is about 68°F. or warmer. She must mate within 5 to 8 days after emergence. If she is not mated, egg laying will begin, but the eggs will be unfertilized and develop into drones.

You can begin your beekeeping adventure in several ways. You can purchase equipment (new or used) and bees with a queen. If you are starting with new equipment, a 3-pound package of bees is suggested. If used or drawn comb is available, you need a 2-pound package. The extra bees help secrete wax and build comb, but tests have shown that if comb is available, there are enough bees in a 2-pound package to care for the queen and feed larvae, and by the end of the season, the package will be as productive as the larger size. If you purchase used equipment, be sure that it is disease-free. Find out why it is for sale: did the previous owner unknowingly have a disease problem, lose the colony, and give up? This could also happen to you if you start with diseased equipment. A knowledgeable beekeeper can accurately diagnose the main bee diseases. Some state laws have regulations against importing, moving, and selling used equipment; some require an official inspection before selling or moving equipment. If you are not certain of your state regulations, contact your local cooperative-extension-service agent, sometimes called a county agent or farm advisor. He represents the U. S. Department of Agriculture and the state land grant university and can help you locate an inspector and inform you on local regulations. If you purchase a full-size hive, observe the same precaution and meet the legal re-

Plans and dimensions of a standardized hive.

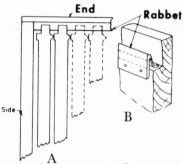

A. Corner of 10-frame body, showing construction and position of frames

B. Part of end of hive body, showing rabbet, which should be made of tin or galvanized iron

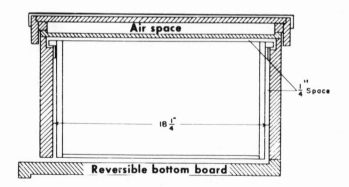

CROSS SECTION OF HIVE BODY AND FRAME

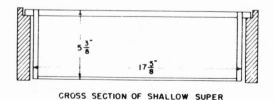

CROSS SECTION OF SHALLOW SUPER

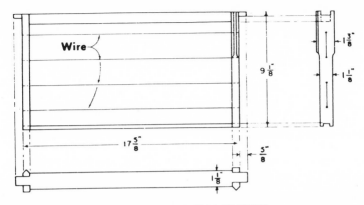

SIDE, END, AND TOP ELEVATION OF FRAME

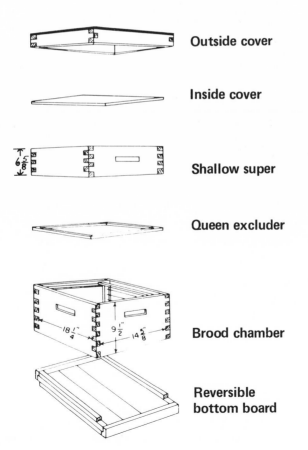

Outside cover

Inside cover

Shallow super

Queen excluder

Brood chamber

Reversible
bottom board

quirements of your area. The inspection requirements for obtaining a permit are sound: at one time disease nearly eliminated the beekeeping industry, so most states have some type of regulation.

If you plan to begin with new equipment, you have several options, such as hobby kits. Suppliers are interested in beginners' success; their business depends on satisfied customers. For obvious reasons equipment is sold in parts and must be assembled. Most suppliers sell component parts of the hive. If you choose this route, plan what you need in advance and order accordingly.

It is possible to make some of your own equipment, especially if you have woodworking equipment and such activities interest you. Several precautions should be noted. If you use commercially prepared equipment as a pattern, be sure to make the parts exactly the same size. As pointed out previously, bees can accurately determine bee space. They will fill in extra space with unwanted comb and seal small spaces with propolis, create needless hard work for you later. Frames should be made to fit commercially prepared foundations. Plans are available from the U. S. Department of Agriculture through your county extension office. Make the component parts to exact dimensions so they fit properly and are interchangeable with commercial equipment. If you decide to terminate your beekeeping enterprise, you will have a good-quality marketable product.

Pine is the most widely used wood, but several other types are satisfactory. The wood should not change dimensions or shape and should be light in weight. If the bottom board is to be in contact with the ground, it should be treated with a wood preservative to prevent decay. The ten-frame Langstroth hive body is widely used in the United States. The two sizes most commonly manufactured are deep (9 5/8") and shallow (6 5/8"). Deep hive bodies are generally used for brood chambers; the shallow bodies, for honey chambers, usually called supers or honey supers. Shallow bodies can be used as brood chambers, but most beekeepers prefer not to use the deep bodies for honey chambers, because they are hard to handle when full of honey. Using shallow bodies for brood chambers and honey supers has the advantage that the frames are interchangeable, which allows for better colony management, but it is slightly more expensive to manufacture. To obtain a 6 5/8" piece, standard 8" stock is required, leaving some waste. To make a 9 5/8" piece, standard 10" stock is employed, with less waste.

Beekeepers using shallow equipment generally allow three or four hive bodies for brood chambers. When deep equipment is used, two hive bodies are allowed. The number of honey supers needed depends on the area in which bees are kept and on frequency of extraction. In areas in which a heavy honey flow can be expected, consider three to five honey supers per colony as a maximum once the colony is established. Some suppliers package several components together and price them cheaper for commercial beekeepers. To save money, you might consider such a unit purchase and divide the material with another hobbyist or beginning beekeeper.

White coveralls are very desirable for personal protection. Gloves and a bee veil are a must. Defending worker bees do not aim when stinging; a sting in the eye can be dangerous and is not worth the risk. It is impossible to avoid crushing bees in manipulating frames and hive bodies. Crushed or stinging bees release the alarm pheromone. Gloves will protect your fingers from these unavoidable stings. A smoker is essential. Decayed wood, dried grass, dead leaves, cotton rags, twine, and many other types of fuel produce satisfactory smoke. Be sure that the fuel does not contain an insecticide: it will kill the bees.

Suppliers offer several types of foundation, some of which are designed for specific purposes. Specialized foundation used for chunk or comb honey is discussed briefly in chapter 4. Through the years the industry has tried to develop synthetic substitutes for foundation and comb. Some of these materials are satisfactory, but no one product or design offers the final solution. A queen excluder is a useful device that you should consider purchasing. Specific uses are discussed later.

If you purchase an established hive, you can begin your operation at any mutually agreeable time. Most beginners prefer to start 1 or 2 months before the main nectar flow. Colonies can be moved at practically any time, but in very cold weather bees may break cluster if they are severely disturbed, which can lead to other problems. Most experienced beekeepers like to install package bees about 2 months before main honey flow, or when the first early-flowering plants appear. In the northern United States (Chicago and above) this occurs around April 1; further north, obviously somewhat later. Early-blooming plants that provide essential pollen are willow, maple, box elder, and dandelion. Packaged bees can be ordered for a specific delivery date. Reputable suppliers can normally meet their commitments within several days. Your equipment should be assembled before the bees arrive and installed soon after arrival. Many suppliers furnish directions on installing package bees and releasing the queen. Literature is also available from the U. S. Department of Agriculture, from many universities through their cooperative-extension service, and from experienced beekeepers.

As with many operations, there are several ways of doing essentially the same thing, but most differences in technique are not significant. There are some problems to avoid. If packaged bees are installed near an existing strong colony, some bees will tend to join the existing colonies. This is especially true if they are installed on a warm, sunny day and allowed to fly soon afterward, even more so if the queen is slow in laying eggs. If you must install packages near full-size hives or or. a bright sunny day, seal the entrance to the hive with green grass. This forces the bees to stay inside and tend the queen. In 2 or 3 days as the grass dries, the bees will open the entrance and begin activity. If you install several packages side by side, bees will sometimes abandon a slow-starting queen and join the stronger package. Ideal package-installing weather is fairly cool (below 55°). Bees will not venture far from their new home in cold weather. Newly installed bees will begin comb building, and in a few days the queen will begin to lay eggs. As with all biological organisms, there are variations: some queens start sooner than others.

Beginners sometimes become apprehensive. They continue to examine newly installed bees and the population continues to decline. This is to be expected. Old bees die, but their replacements have not developed. It takes 21 days for a new bee to emerge, and as a rule the population is at its lowest level at this time. In 30 days the population will reach its original level and thereafter will increase, depending on the queen's egg-laying capacity and the availability of pollen and nectar.

Supplementary Sugar and Pollen Feeding

Sometimes no nectar and/or pollen are available. Bees are biologically successful because they are capable of storing for such periods. Man is the honeybee's greatest enemy: more colonies are lost because of mismanagement — by taking too much honey — than from other causes. Shipping and installing packaged bees is an abrupt change. Bees can successfully survive this disruption if they have sufficient food — that is, honey and pollen. Through the years supplementary diets have been developed. Cane or beet sugar (sucrose) can replace honey. Bees actually convert it into honey, but it is usually consumed before it becomes a true honey. The conventional manner in which sugar is fed is to dissolve it in hot water. A thick syrup is made by dissolving 2 pounds of sugar in 1 pound (pint) of water. The water must be heated in order to dissolve the sugar. If the sugar is more concentrated, some will crystallize with cooling and be wasted. Some beekeepers prefer to feed a more dilute mixture — 1 pound sugar to 1

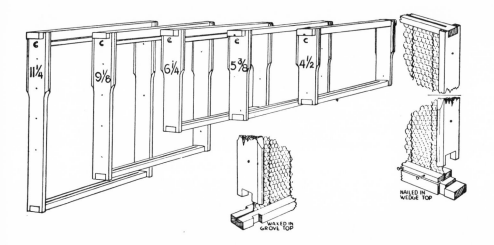

Most popular size hive bodies are 9 5/8" and 6 5/8" but larger and smaller ones are also available.

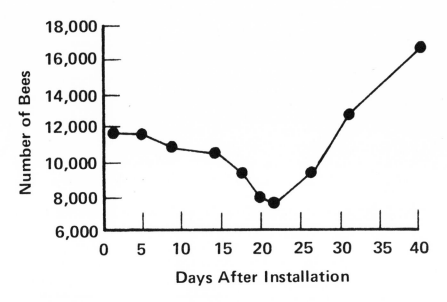

Population changes in a 3-pound package of bees. Since no bees emerge to replace those that die of old age, the population declines for 21 days after the queen lays her first egg.

pint water. Bees will store this material. The point to consider is that while bees need water, sugar is the energy food. If they are fed a more dilute solution, they have to evaporate the extra water. Dry sugar can be fed, but the bees require moisture or water to dissolve the crystals. If they are fed in warm weather, they will bring water to the hive. At low temperatures, however, bees will starve rather than venture out.

There are many different ways to feed sugar syrup to bees. Each technique has its own advantages and disadvantages. Commercially manufactured feeders are available, or you can make your own. A container is required to hold the sugar syrup, and a dispensing device is needed. It can be holes in an inverted can, such as a metal coffee can. Bees will suck the syrup out through a few small holes. This device has one distinct advantage: bees can be fed at relatively low temperatures compared with other devices. The heat from the cluster will rise and keep the container warm. Bees tend to cluster around the small holes and feed even at fairly low temperatures. Sugar syrup can be poured directly into empty combs, which are then placed in the colony. This is not practical for beginners who do not have combs, but if you have the equipment, it is highly satisfactory. Open containers inside the hive can be used, but bees need a float so that they do not drown in the syrup. Some beekeepers place sugar syrup in open containers outdoors. Bees will carry it into the hive, but you never know how many of your neighbor's bees you are also feeding. You need not worry about overfeeding: bees use only what they require and store the surplus for future use. Substitutes for sugar have been examined. Honey is part glucose and bees can digest glucose, yet they will not feed on it. Corn syrup is pure glucose; some feel that bees will not feed on corn syrup because it is not sweet enough — glucose is about 80% as sweet as sugar. Bees cannot digest lactose (milk sugar), as they do not have suitable enzymes.

All living creatures require protein, especially when they are growing. Larval bees require more protein than adults. If packages are installed when early-spring plants are in bloom, workers will gather sufficient pollen to rear brood. Inclement weather does occur on occasion and workers will continue to rear brood at the expense of their own body tissue if pollen is not available, but brood rearing stops abruptly once their body reserves are exhausted. This happens in about 2 weeks. After years of research in private industry and government laboratories, pollen supplements and substitutes have been developed. While there is no true replacement for pollen, these substances extend the available pollen supply and/or delay depletion of body protein reserves. A true replacement would mean that brood rearing could continue indefinitely without additional pollen: the substitutes and supplements allow brood rearing to continue un-interrupted for a limited period of time. Many products have been evaluated as possible protein supplements, but only two are satisfactory: brewer's yeast and soy flour manufactured by the expeller process. While bees can be forced to consume other products, these are useless unless they are digested.

After bees are installed, with workers building comb and the queen laying eggs, you need only check the colony periodically. Some prefer to check weekly during the spring buildup; others examine them every 2 weeks. Swarming is not a problem in colonies started with packages, because the population is not too large. It usually occurs in older, well-established colonies.

Seasonal Management Techniques

Producing honey can be compared to many other types of businesses. Capital, labor, and management skills are required to produce a product that someone else needs or wants. By skillfully applying known information about the biology, life cycle, and behavior of the honeybee, a good manager can fully utilize the productive capacity of the colony. Since bees are biological organisms, variations should be expected. Modern technology has reduced some of the seasonal and between-seasons variations and risks in the production of many agricultural products. The beekeeping enterprise, however, is at the mercy of nature to a greater degree than other operations. The beekeeper has no control over the climate that affects the activities of bees and/or the nectar secretions of plants. Some commercial beekeepers move bees to take advantage of either climate or nectar sources. The basic objective of colony management is to coordinate and synchronize colony development with plant resources. Honey production is directly correlated with colony size — the larger the colony, the greater the honey crop. If the nectar-flow period is relatively short, it is critical that the colony size be precisely synchronized; on the other hand, if honey production extends over a longer period of time, synchronization is less critical.

The amount of labor invested in a beekeeping operation will vary considerably. The hobbyist may overlook labor costs entirely, whereas a part-time or full-time commercial enterprise may have to make decisions based on labor costs. The skillful manager can pick and choose ideas and operations suitable to his interests, needs, and size of operation. In a business enterprise the fiscal year may not synchronize with the calendar year for logical reasons. The same is true with beekeeping. The new year actually begins in October or November. To simplify the discussion, it starts with January on the assumption that the colonies were adequately prepared in the fall. The monthly operations are generally described for the northern United States (Chicago and above). Further north the season would be somewhat later; further south, earlier. In the far south some honeybee activities take place all year. Honey production is governed to a degree by the availability of nectar-producing plants rather than by the season.

January

Low temperatures at this time of year are of little or no concern to the beekeeper if the colony was properly prepared in the fall. Bees easily withstand outside temperatures of -20°F. to -30°F. (-28°C. to -35°C.). The cluster will congregate in the upper portion of the hive where it is somewhat warmer. Observations over many years indicate that larger colonies are better able to survive cold weather than small ones. Some beekeepers feel that if a large colony is allowed to survive through the winter, it will consume greater quantities of honey, use up the supply, and starve in spring. By periodically weighing colonies it was learned that honey consumption is quite modest until early February. Bees do consume honey at all times, but during brood rearing the greatest quantities are used. Bees in a cluster act as an insulator: the larger the cluster, the greater the insulation value and the less heat loss. The queen is located inside this cluster. She will begin to lay eggs in cells that were cleared of pollen and honey. Bees produced at this time are very important to the colony, as they will be involved in brood rearing in February and March.

A prolonged cold spell of 1 or 2 weeks can cause problems. A tightly confined cluster can consume all the honey and starve, especially if it is small to moderate in size. On the other hand, a large cluster will undoubtedly cover a large enough area of honey to survive. Low temperatures for 1 or 2 days followed by a warmer period present no problem, because during the warmer period the cluster has a chance to expand and move to a section of the comb containing honey. Fluctuating winter temperatures are highly desirable. In areas with moderate climates — O°F. to 20°F. (-18°C. to -7°C.) — winter survival presents no serious challenge to a healthy colony. Bees do cluster, but the cluster is not as tight as it would be at lower temperatures, and during the warmer part of the day the bees can move on to more honey. The rate of egg laying in January is directly related to the availability of pollen and the size of the cluster. The queen will lay eggs only at a rate at which the workers can keep them warm. Egg laying will stop in about 2 weeks if pollen is lacking.

An inspection on a warm day — above 32°F. (0°C.) — in January can often save a colony. Carefully and slowly remove the cover and the inner cover. Blow some smoke over the cluster to prevent bees from flying out. Being careful not to disturb the cluster, examine combs to either side of the cluster. If they are empty, replace them with a comb of honey, preferably one that also contains some pollen. This can usually be done by switching a comb near the side of the colony with one next to the cluster. It is not unusual for a colony to starve because honey was one or two combs away. In moderate climates this operation is less critical, because bees will readily move to honey or move honey near the brood area as needed.

On a warm, sunny day in January or February bees will fly out of the top entrance to void feces; some may not return. If the ground is snow-covered, dead bees are conspicuous and may cause considerable apprehension for the beginning beekeeper. This phenomenon of bees leaving the hive and not returning is normal. Bees tend to die (from old age) outside the hive. At this time of the year the fall bees are dying and being replaced by the new ones. If the weather is cold — 32°F. (0°C.) or colder — these bees usually die inside the hive. As the weather continues to warm up, workers will haul out the bees that died inside during cold weather. A pile of dead bees around the hive in late winter or early spring is desirable: it means that the hive has survived and that the bees are doing some cleaning.

February

The February inspection is very important. The tempo of brood rearing has increased, and honey is disappearing quite regularly. Colony inspections can be made quite rapidly if an organized system is used. Always wear protective clothing and a veil and blow smoke into the opening. Avoid pounding on the hive. Lift off the cover gently, then blow some smoke into the hole on the inner cover. Pry off the inner cover with the hive tool and blow some more smoke over the frames to keep bees from flying upward. If it is necessary to remove a comb, always start with the end comb or the one next to it. By using this technique there is less likelihood of disturbing the cluster and/or injuring the queen, since there is less chance that she is in that area. If one frame is removed, the others can be easily pried loose and lifted for inspection if required. In cold weather the less you disturb the cluster, the better. Any heat lost through hive manipulation is replaced at the expense of honey and extra effort. Especially in colder climates it may be necessary to move frames of honey next to the cluster. In warmer areas bees will move around and transport honey to needed areas.

It takes considerable heat to elevate the temperature of cold honey to that maintained inside the cluster. Do not disturb the cluster by placing a frame of cold honey in the center: instead place the frame *next* to the cluster, as the bees will expand to move over it gradually. If honey is lacking, sugar syrup poured directly into the combs provides an acceptable substitute. Sugar syrup is made by dissolving 2 pounds of sugar in 1 pint of water. Most colonies starve during February and March. Man continues to be the honeybee's worst enemy by removing too much honey.

March

Operations performed during March determine to some extent the size of the honey crop produced that summer. Observations over a period of many years indicate that the limiting factor in March and April brood rearing is a lack of pollen or improper location of pollen in the hive. Bees will move honey within the hive, but pollen is eaten by brood-rearing workers directly from the cells. If pollen is in the lower part of the hive and the cluster is in the upper part, it is of no value to the bees. Honey consumption dramatically increases as the queen continues to lay eggs. It is estimated that one cell of honey and one cell of pollen is needed to produce one adult bee. During this period colonies should be periodically checked to be certain that suffcent honey and/or sugar syrup is available. Some beekeepers find their strongest colony dead at this time. The largest colony uses the most honey: unless adequate supplies are available, they are the first to run out of honey and starve.

There are many advantages to building a large colony early, even though it requires more careful mangagement, especially in terms of swarming. Honey production is directly correlated with colony size: a large colony produces more honey more efficiently. Large colonies may also be divided into two or even three colonies, or the beekeeper may wish to sell some bees and realize additional income, so there are a number of valid reasons to develop a large colony. Tests prove beyond doubt that supplementary protein feeding greatly enhances honeybee brood rearing, which in turn produces strong colonies. A strong colony can collect more pollen and nectar from early-flowering plants such as willow, elm, and maple, which in turn further stimulates brood rearing. The shortage of pollen in the normal colony in a temperate climate restricts brood rearing. This problem can be par-

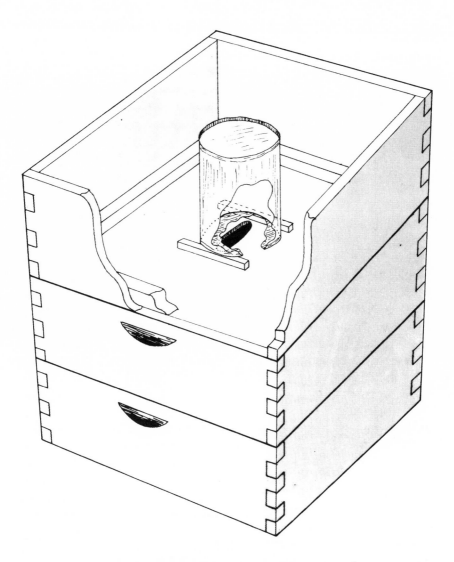

Feeder pail inverted over the inner cover for supplementary feeding.

tially corrected by feeding supplementary material containing essential nutrients that bees will consume and digest. Brewer's yeast and expeller-processed soy flour are two such products. These materials do not replace pollen but do serve as effective extenders, thus providing the necessary protein for brood rearing to continue uninterrupted until adequate pollen is available.

The beekeeper has several management options: (1) Pollen can be trapped from a few colonies during the late spring or early summer, stored, and extended the following year with soy flour or brewer's yeast. (2) Brewer's yeast and soy flour can be purchased. (3) Ready-to-use mixtures can be purchased from bee-equipment suppliers. These mixtures consist of dry powders, which are mixed with a sugar syrup, formed into a dough patty, and placed over the cluster on top of the frames. The patty is usually covered with a piece of wax paper to keep it from drying out. Bees will consume up to 2 pounds of this material in a week. As soon as natural pollen is available, they will stop eating the mixture. While soy flour and/or brewer's yeast alone is not consumed as readily as mixtures containing pollen, these materials do effectively extend the pollen in the hive.

Bee diseases are discussed in detail in chapter 6, but one intestinal parasite (*Nosema apis*), a protozoan, is endemic in practically all bees in the United States and probably in other countries. If it is left uncontrolled, especially in northern temperate climates, it can drastically reduce colony vigor. It shortens the life of the adult bee and causes a type of dysentery. Nosema spreads rapidly in the colony especially during periods when bees cannot fly outside. In late spring and summer this disease tends to somewhat eliminate itself. Both healthy and sick bees have a strong urge to leave the hive if they are dying or about to die. This behavior benefits the colony by not exposing healthy bees to disease-producing germs. During November, December and January (especially in northern areas) bees usually die in the colony. Those infected with nosema unavoidably soil the comb and frames, and the infectious organism remains alive in the feces. As the colony expands in size during March, young workers clean and polish the soiled comb to provide the queen with cells in which to lay eggs. During this process these young workers become infected. At this time it is highly advantageous to feed the bees sugar syrup in which a drug called fumagillian is suspended. As bees ingest the syrup and feed it to others, they treat themselves. Tests under controlled conditions demonstrate conclusively that fumagillian protects bees and that protected colonies outproduce untreated colonies. Some beekeepers mix fumagillian with the pollen supplements or substitutes. This is satisfactory, especially if the colony has sufficient honey in the comb and needs no additional sugar-syrup feeding. Extensive tests have also shown that if fumagillian is supplied early in spring before the nectar flow, no residue will be found in the honey. The use of drugs and medications is regulated by state and federal agencies. One requirement imposed on suppliers is that adequate directions by provided to the user. It is the beekeeper's responsibility to read and follow instructions accurately.

If the supply of honey is low, sugar syrup must be fed. The conventional heavy syrup is made by dissolving 2 pounds of sugar in 1 pint (pound) of water. There are many different ways to feed this mixture. Bees will store it in combs just as they do nectar. During cold weather bees will not leave the cluster to feed. An inverted container such as a used coffee can with several holes in the lid has decided advantages as a feeder pail. The container can be placed directly over the cluster, and heat from the cluster warms the sugar syrup. In an emergency or in very cold weather warm sugar syrup can also be fed. There are other feeding devices, such as the division board and the Boardman feeder. Dry sugar has been used successfully as emergency food. Some beekeepers keep dry sugar in their behicle when making routine inspections in the spring. If they find a hive that needs food immediately, a handful or two of sugar sometimes means the difference between survival colony and starvation. When the temperature is in the 40- to 50°F. range, the hives can be inspected thoroughly. The hive bodies should be lifted off the bottom board, and dead bees and debris removed. Sometimes the bottom hive body is completely empty. The chamber can then be removed, and hive bodies containing bees placed directly on the bottom board. If the bottom brood chamber contains honey and/or pollen, it can be left in place. Bees will move the honey up and store it around the brood area. If the chamber is full or nearly full of honey and pollen, it can be placed on top of the colony. The cluster will gradually eat its way into the full combs, converting honey and pollen into bees.

April

April is the most active month for beekeepers. On warm days in northern areas bees begin to work early-blooming plants. These plants often produce large quantities of very important pollen and some nectar. It is important to check hives periodically to ensure that they have sufficient space to store pollen and nectar. Experienced beekeepers can make routine inspections quite rapidly. Armed with a smoker and a hive tool and wearing protective clothing, the hive is approached from the side or the rear. Smoke is blown into the entrance; the hive cover is slowly removed; and an additional puff or two of smoke is blown into the hole in the inner cover. The inner cover is slowly removed, and more smoke is blown over the frames. If bees are in the uppermost chamber near the top, some frames are carefully removed and examined. It is best to pry loose the outside (or next to the outside) frame. Take it completely out of the hive and set it on one end against the hive. The reason for removing an outside frame first is that there is less possibility of injuring the queen. There is a greater likelihood that she will be in the center of the colony, especially at this time of the year. The other frames can be pried loose and carefully lifted out for examination.

Many beekeepers like to divide or split their colonies in April. It is common practice to overwinter only strong or large colonies and divide, split, or increase in the spring. It is important that this colony be fed supplementary protein during the previous month. Dividing a small colony is not suggested, because the number of bees may be below the critical mass. While it may survive and prosper, it might well do better as one strong colony. One strong colony will produce more honey than two weak ones. To divide a colony, it is necessary to order in advance a queen from the south. While bees will produce a queen at any time, it is important to have favorable temperatures and other climatic conditions so that she can fly out and mate. In northern areas the weather at this time of the year is unpredictable, so even if a queen is produced, she may not mate and become a drone layer.

Assume, for example, that you have a strong colony in two deep (9 5/8") or four shallow (6 5/8") hive bodies or brood chambers and that it is to be divided into two colonies, each in one deep or two shallow chambers. Have a mated queen available and examine each chamber carefully: each should contain honey and pollen. It may be necessary to move individual frames from one chamber to the other. Combs containing brood should be equally divided, though this may not be critical, as colonies can be balanced later. The chamber containing the old queen should be placed on the bottom board and covered. The inner cover works quite well, provided that the hole in the center is covered with something like a piece of plywood. The second hive body should have an auger-hole entrance. Some beekeepers prefer to face this entrance in the opposite direction from the bottom one, but this is not essential. You now have two colonies — one on top of the other. The bottom one contains a queen; the top one is queenless. Your new queen, still in her shipping cage, is now introduced into the top colony. One end of the cage is sealed with a candy plug. With a nail or other sharp object, make a hole in the plug. The cage should be placed between two frames with the candy plug facing upward — bees will eat away the candy plug and release the queen. It is important that the opening in the queen's cage face upward: if it were not upward and one or more of the attending workers in the cage should die, the bodies would block the entrance and prevent the queen from leaving the bee cage. As the bees eat away the cand plug, the colony will become adjusted to their new queen and accept her. She is normally released in 2 or 3 days and often starts laying in 5 days. Most beekeepers routinely reinspect their new colonies in 3 or 4 days. If the queen has not yet been released, she is taken out by removing the screen cover of the cage. Some commercial beekeepers order 10% to 15% extra queens. If for some reason the colony does not accept the new queen, another attempt is made at requeening.

As with many other beekeeping operations, there are several ways to divide colonies. Some beekeepers set the equipment for their new hive 2' or 3' from the old one. About half of the combs containing honey, pollen, brood, and attached bees are placed in the new hive, along with the old queen. The new queen is then released in the old hive by the method just described. The reason for taking the old queen to the new location is that bees tend to return to the old hive and to abandon a newly established colony. If their queen is present, they will not abandon her in the new location. Field bees will continue to return to the old location and accept their new queen. Another way to divide a colony is to move the old colony about 2' or 3' from the original location and to place the equipment for the new colony on the site of the old colony. About half of the combs of pollen, honey, brood and attached bees are transferred back to the site of the old colony. The old queen is left in the old colony and the new queen introduced into the new one, again using the slow-release method.

Beginning beekeepers sometimes have difficulty in locating an old queen, especially if she is unmarked. Commercial beekeepers do not want to take the time to locate the old queen but still want to divide the colonies. Understanding the biology of the colony allows them to do this quite easily and successfully. Approximately 3 to 4 days before the new queens are due to arrive, each colony is inspected. The honey and pollen are adjusted so that there are equal amounts in each brood chamber. The inner cover, with the center hole closed, is inserted between the brood chambers. Both brood chambers have entrances, either the conventional entrance by the bottom board or an auger hole. At any time after 3 days the colonies are ready for rapid requeening. All that is necessary is to look for eggs: if the cells contain eggs, the queen is present in that brood chamber. This also means that the other chamber is queenless, so the new queen is introduced, using the slow-release technique previously described. It is very important, however, to destroy all beginning queen cells. If they are not destroyed, the colony will not accept the new queen but continue to raise its own. More than likely the new queen will be killed as soon as she emerges from her cage by workers loyal to the developing queen.

When you are working with living material, it is unrealistic to expect 100% survival. Many beekeepers order extra queens, which can be kept in the mailing cages for several days at room temperature. They will benefit from a drop or two of water placed daily on the screen cover. If the queens are to be held for a prolonged period, each should be placed in a nucleus — a queen, some brood, and attending bees. Queens can be held in nuclei practically all season. Nuclei can be formed in many different ways. All that is required is a compartment for a queen, workers, brood, some honey or sugar syrup, and pollen. Some beekeepers form nuclei containers by dividing a standard hive body into three separate compartments. This can be done by nailing two pieces of plywood parallel to the direction in which the frames fit in the hive body. An auger hole is required in each compartment. The center-compartment opening usually faces forward, while the two side-compartment openings are to each side. The bottom is covered with a solid sheet of plywood, and the top covers must fit tightly so that bees cannot pass from one compartment to the other. Two or three standard frames will fit in each compartment, and the bees, with their queen, will maintain themselves in these separate compartments without difficulty. When a queenless colony is discovered, the frames, brood, attending bees, and queen are introduced into the colony. Some beekeepers construct individual boxes to hold three or four standard frames as nuclei. The same basic principle is employed: the queen and a small force of workers are held in reserve.

May

Management decisions made during May greatly influence honey production for the coming year. Many decisions should be made on the basis of what is happening inside the colony rather than attempting to follow a rigid calendar schedule. A few basic rules on colony behavior should be understood. (1) The queen tends to work upward in the colony. (2) Swarming represents a distinct loss of honey production and should be prevented. (3) A crowded colony reduces foraging, so bees should be given ample room. Colonies started from packages in the spring as a rule will not swarm. Swarming is also a lesser problem with colonies divided in the spring, than with undivided colonies. If colonies were fed supplementary protein in early spring,

Mechanically handling colonies. Such devices relieve the burden of heavy lifting.

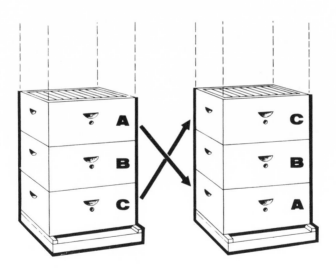

Reversing the brood chamber. Moving the top chamber to a lower position places the queen, eggs, and young brood on the bottom of the hive. The lower chamber, with empty comb and emerging brood, gives the queen enough space in the brood chamber to continue laying.

the swarming instinct will be very strong, primarily because of colony size.

Colonies should be inspected on a weekly, 10-day, or 2-week schedule. Each procedure has its merits; labor requirements are obviously greater as the inspection frequency increases. Inspection need not be time-consuming if you understand some basic principles of bee behavior and look for clues as to what is going on inside the hive. Experienced beekeepers can often tell if a hive is queenless as soon as the cover is removed; a colony about to swarm is easily noted. On a routine inspection you should carefully lift out a comb and look for eggs. If eggs are present, then the queen is present and laying in the upper part of the hive. There is no valid reason to spend time looking for the queen. Combs taken out of the chamber should be replaced in the same position. With the hive tool, pry the hive bodies from the bottom board and tip them back. Beginning beekeepers are reluctant to do this, but it does not harm the bees and is quite easily done, especially on fairly level ground. On their back side, the hive bodies can be separated with the hive tool. The underside of each brood chamber should be examined for swarm cells. Bees build cuplike cells on the underside of the comb, which are of no concern unless one contains an egg or developing larva. If left undisturbed it will become a queen. If an egg or larva is found in any swarm cell, examine the hive very carefully. All swarm cells must be found and destroyed. If one is missed, the hive will swarm, possibly within a week.

If the top chamber contains eggs and developing larvae, place it on the bottom board. Young workers will move down to the lower position and continue feeding brood. The chamber that was on the bottom will contain either empty cells or mature larvae. Place it above the one containing eggs. This process is called reversing. If three shallow chambers are involved instead of two, reverse the top and bottom chambers around the middle one. If the colony is not reversed, the swarm cells would continue to develop, the workers would slow their foraging and prepare to swarm, and on a bright, sunny day they would leave the hive.

If no eggs are found on a routine inspection, look for supercedure cells, queen cells located on the face of the comb rather than along the bottom of the frame. The presence of these cells suggests that the queen has died or is missing. Two options are now available: (1) to requeen with an extra queen or (2) to allow the colony to requeen itself. If the colony is allowed to requeen itself, all that you need to do is to inspect the colony in about 2 weeks to ensure that the queen is not a drone layer. If requeening is intended, it is necessary to destroy all developing queen larvae. If extra laying queens are maintained in nuclei, remove several combs from the queenless hive and replace them with combs, brood, attached bees, and the queen from the nucleus. Some beekeepers make sure that the queen is somewhere in the center of the nucleus. If the nectar flow is on and it is a sunny day, the queenless colony readily accepts the new queen and other bees. If requeening must be done under less favorable conditions, the chance of successful requeening is greater if the bees are sprayed with a light sugar syrup, made by dissolving 1 pound of sugar in 1 pint (pound) of water. If a queen without a nucleus of brood and bees is to be introduced, the slow-release technique must be used. Some beekeepers will divide the queenless colony between several other colonies. As mentioned previously, a strong colony will produce more honey than two weak ones.

June and July

In many areas of the country nectar flow begins in June. Honey supers should be ready, as crowding tends to reduce foraging. Colonies should be inspected on a weekly, 10-day, or 2-week schedule.

The queen excluder is a device that allows workers to pass through but not queens or drones. It has a definite place in colony management, but its use in other areas is questionable and sometimes controversial. Its purpose is to confine or restrict the queen to a specific area and not to let her wander into honey supers where she will lay eggs. Workers must do extra work to crawl through an excluder. Extra work means extra energy expended or at least a slowdown of activity. For this reason some beekeepers believe that a queen excluder is also a honey excluder. Timely reversal of brood chambers tends to keep the queen in her brood nest. A queen excluder is required in a two-queen colony system. It is highly advantageous to install it while using wet supers, which are discussed later in this section.

In addition to being a manager of colonies, a beekeeper is also a bee breeder, not in the sense that he produces queens but that he selects, maintains, and propagates colonies of bees with desirable traits. At each inspection you should note whether the bees are highly defensive or gentle. Brood patterns should be noted: the eggs and larvae should be in a full, consistent pattern — that is, all cells should contain eggs or larvae. A pattern with many skips indicates that the queen might be laying some nonviable eggs. A large queen is preferable to a small one. All these traits are inherited, and, following the tradition of successful livestock breeders, bees with undesirable traits should be eliminated from the system. All that the beekeeper has to do is to replace the queen. Auger holes in the brood chamber tend to increase worker efficiency and provide additional ventilation. A hole about 1" in diameter located below the hand grip is satisfactory. This opening also serves a vital function in successful winter management. Nosema-infected bees can easily leave the hive, thus reducing the possibility of infecting other bees.

When supers are used for the first time, the frames will contain only foundation. It is important that each hive body contain its full complement of frames. Bees will build comb on foundation, but if the space between foundations is too large, they will build comb in all directions. Bees are reluctant to venture into a super containing only frames of foundation. They can be induced or baited upward by exchanging one or two combs (preferably of unsealed honey) from the brood chamber with one or two frames with foundation. Once bees begin to build comb, others will venture upward. Do not add frames with foundation on hives unless bees are being fed sugar syrup or a nectar flow is underway. Bees like to chew holes in the wax foundation, which reduces the possibility of obtaining good combs. The best combs are produced in supers during nectar flow. Combs built in the brood chamber are usually not attached around all edges of the frame, and sometimes holes are chewed out of the corners. Once combs are constructed, some beekeepers use nine frames per standard ten-frame hive body. It is important that they be equally spaced. Bees fill the cells to the top with nectar. If additional space is available between combs, other bees will build extensions on existing cells, which results in very thick combs that are easier to uncap when extracting honey.

Not all bees return to their own hive: the phenomenon of drift has been observed for years. In a row of colonies, all somewhat identical and all facing in the same direction, the two outside colonies generally have more bees. Beekeepers have attempted to correct the problem with varying degrees of success: painting designs, painting hives different colors, and facing them in different directions have all been tried. Some insist that their system works. The simplest way to correct the problem is to periodically remove one or two frames of sealed brood from the strong colony and to place it in a weaker colony. As the bees emerge, they will stay in their new home, balancing the colonies.

As nectar comes into the hive, it contains from 50% to 80% moisture. Nectar is deposited above and to the sides of the brood nest. Colony inspection during the honey flow is important. while the danger of swarming has passed, the beekeeper should prevent a colony from becoming honey-bound. This condition exists when the brood nest is completely surrounded by fully sealed combs. The bees apparently believe that they have enough honey and reduce foraging activities. The first supers should be added just before the nectar flow begins. Soon afterward additional supers can be added on top of the first super. This procedure is called top supering. If the first super is nearly full of honey on the next inspection, it is desirable to lift it off, replace it with one or two empty supers, and set the nearly full super on top of the empty ones. This procedure is called bottom supering. While it requires heavy lifting, it does prevent colonies from becoming honey-bound. Top supering works best early in the season and when nectar is coming in quite rapidly. Bottom supering tends to increase colony efficiency. Bees seal honey only after moisture is reduced to levels that assure safe storage. A rule followed by many beekeepers is to remove and extract combs once they are 80% sealed or capped. The rate at which honey is capped varies considerably from season to season, depending on such factors as moisture content of nectar and relative humidity of the atmosphere. There is no reason to leave honey in the hive once the combs are capped. In fact some beekeepers believe that the sooner the honey is removed and extracted, the less opportunity it has to pick up an objectionable odor. Some beekeepers like to keep one colony on a scale and record weight changes during the nectar flow. It is important to weigh the colony at about the same time each day. Colonies usually gain from 1 to 15 pounds per day, but they might actually lose weight if there is little or no foraging but other bees are evaporating water from the nectar within the hive.

The size of your operation determines which technique should be used to remove honey. The hobbyist with two or three hives will probably remove individual combs as they are sealed. Bees can be removed with a soft-bristled brush, and the comb placed in a covered empty hive body and taken in for extraction. Some beekeepers remove only full supers. After deciding that the super is 80% capped, it is pried loose with a hive tool and rotated 180°. Turning the super breaks the comb (usually built between supers). Bees will clean up the dripping honey from these combs, and the super is cleaner to handle. Bees can be removed from honey supers with bee blowers, bee escapes, or repellants. The blower is probably the most efficient: a stream of air is used to dislodge bees from supers. While it is fast and effective, the investment is considerable unless the equipment has other uses or the operation is large enough to justify the cost. The bee escape is a device that fits into the standard hole in the inner cover or a board of similar construction. The sealed super is placed on top of the escape board or inner cover with the escape. The escape allows bees to move out of the super but not back up into it. It is a one-way bee-traffic regulator. Bee escapes should be installed about 24 hours before super removal.

Equipment must be bee-tight, or bees will job the unprotected super of honey. Escapes should not be used during a hot day, since bees cannot cool the super, and if the temperature is too warm, the combs will melt. Repellents can be used to drive bees down into the brood chamber. In order to use them effectively, the beekeeper must pay attention to details. Selection is made on the basis of temperature. The most common repellents are propionic anhydride, benzaldehyde, and glacial acetic acid. These products are placed on absorbent pads above the colony. Repellents combined with smoke will drive the bees out of the super.

Honey should be extracted from combs as soon as possible. The combs can then be replaced on hives. Commercial beekeepers may delay extraction for business reasons. Combs that have just been extracted are called wet combs. If the apiary is located close to the honey-extracting house, special care should be taken that these combs are not exposed to bees. They should be kept covered until they are placed on the hive. Supers containing wet combs should be placed above honey when they are returned to the hive. These combs are highly attractive to bees, and workers tend to move to them quite rapidly. The queen often moves along with the workers and begins to lay eggs in the honey super. To prevent this from happening, many beekeepers place a queen excluder between the brood nest and the wet combs. Supers with wet combs can also be placed above the inner cover. Workers will move up through the opening and clean the combs, but the queen will seldom venture up that far.

August and September

In many parts of the country the nectar flow slows to a mere trickle in August and September, but enough usually comes in for the colony to maintain a constant weight. In other areas the late nectar flow may produce a significant crop of honey. Late-flowering plants like goldenrod and aster can yield a good crop of honey if conditions are favorable. As a general rule late-blooming plants produce darker honey. Some beekeepers like to segregate this type from the earlier product. This is a highly individualistic matter, and commercial beekeepers may segregate or not, depending on their particular market demands.

In September especially you should think about which colonies, if any, are to be eliminated. Colony reduction can be done in several ways. The simplest is to divide the hive bodies between three or four colonies to be retained. The brood chamber can be placed over a sheet of newspaper on top of an existing colony. Bees will peacefully unite as they chew away the newspaper. If the colony to be destroyed has an undesirable queen, it is best to find and kill her. If the two queens are allowed to fight it out, the less desirable one from the beekeeper's point of view might survive. Some beekeepers locate the queen in the colony they wish to eliminate and confine her in a screen cage. She will keep the colony working if pollen and nectar are available, yet eggs will not be laid. After 3 weeks, all brood will have emerged, and the adult bees can be shaken out and forced to join other colonies. With this technique the combs are free of brood and can be put in storage for next year.

Some beekeepers in northern climates use the latter techniques for commercial honey production. The honey and pollen in combs are stored until the following spring, when package bees are installed. The newly installed packages can be started somewhat earlier and may require less feed. Beekeepers sometimes have partially filled supers of uncapped honey. Bees can be forced to transport their honey into the brood nest in two different ways. (1) Partly filled supers can be placed above the inner cover. The caps should be broken with a heavy brush or hive tool. Bees will come up through the opening in the inner cover and move the honey down to the brood nest. (2) The super can be placed on the bottom of the hive, while honey below the brood nest is moved upward. The latter technique works extremely well, although it is hard work to lift off the hive, place it on top of the partially filled super and, in about 2 weeks, do the same to remove it.

In manipulating hives and handling honey at this time of year you should be alert to the possibility of robbing. Bees are opportunists and will carry honey or nectar from any source back to the hive. The more concentrated the sugar, the more attractive it is to bees. The odor of honey is easily detected by bees and readily hauled back. The message of available honey is quickly spread to other bees in the hive, and robbing is underway. Experienced beekeepers can readily spot robbing workers. They usually dart around very rapidly. If they land on a comb, they quickly go for an open cell and start to engorge. In flight their legs usually hang downward. Robbers apparently have a good memory, since they will return to the scene days after their source has been removed and continue to search. They will also attempt to steal honey from other colonies and will overpower a weak bee or a colony that cannot defend itself. A colony cannot defend itself against robbers if the entrances are too large or if the hive bodies do not fit tightly. Robbing, once started, is difficult to stop, so good beekeepers make an extra effort to prevent robbing rather than trying to control it.

Robbing can be prevented or discouraged if the following points are kept in mind when little or no nectar is coming in. (1) Never expose honey outside the hive. Materials such as broken combs and scrapings that might contain some honey should be placed in a covered container. (2) Do not expose freshly removed equipment such as hive covers, bodies, and frames. (3) Colonies located near buildings where honey is handled should not be manipulated while honey is being extracted. (4) Do not expose extracted supers and other equipment to bees. (5) Vehicles hauling honey from the apiary should be unloaded in a bee-tight building. (6) Manipulate colonies as quickly as possible during periods in which robbing could occur. (7) Honey equipment that must be exposed during this period should be covered with a wet cloth. Wet burlap is a good covering.

Robbing can be stopped in the following ways. (1) Discontinue colony manipulation and reduce the entrances to all hives. This procedure will help each colony to defend itself. (2) If possible, kill or prevent the first few robbers from returning to the hive. They will stimulate others to begin robbing.

October, November, and December

Consider October as the beginning of the beekeeping year. In most parts of the country the nectar flow is over, and egg laying is greatly reduced. This is a good time to cull the weak or poorly performing colonies. It is far better to eliminate weak colonies now than to attempt to carry them through the winter and lose them later.

Many commercial beekeepers routinely requeen in the fall. It is a little more difficult in the fall than in the spring, since the colony has more bees and they are usually more defensive, but some beekeepers feel that the benefits outweigh the difficulties. In northern areas some beekeepers raise their own queens during late summer or purchase them more cheaply than in spring. There are varying opinions on how often you should requeen. Some beekeepers requeen annually, either in spring or fall; others keep the same queen two autumns and one summer. The harder the queen works, the faster she is exploited.

Many beekeepers inspect and rearrange the colony before feeding. Since the cluster is going to be in the top of the hive in midwinter, it is important, especially in northern areas, that some combs in the top chamber contain pollen. If all the pollen is in the lower chamber, it is wise to switch a few frames now rather than in spring. Pollen in the lower chambers would not be available to the workers during cold weather when they need it most. Some commercial beekeepers routinely weigh each colony in late fall and add sufficient sugar syrup (2 parts sugar to 1 part water) to bring each colony up to its minimum weight (1 gallon of sugar syrup adds 7 pounds). Do not be afraid to overfeed sugar syrup: bees, unlike many other types of animals, eat only what they need and store the extra food. As winter approaches in the northern United States the colony should have up to 90 pounds of honey; further south, 50 to 60 pounds are sufficient. Fall-feeding techniques are essentially the same as for spring. Inverted containers with holes, division board feeders, and other devices work well. The sugar syrup will be carried down into and around the brood nest. Be sure to include fumagillian in fall feeding. Although nosema infection is low at this time of year, the drug is stored in the sugar syrup and will protect bees during the midwinter season when the stored sugar syrup is consumed.

Winter losses have been a major concern to beekeepers for many years. In northern areas these losses are associated with cold weather, and people have attempted to correct the situation in different ways. Some have moved colonies into dark basements; others have wrapped them with insulating materials such as corn stalks, straw, burlap sacks, or heavy paper. Suppliers now provide plastic or cardboard insulators for protection. There are differences of opinion as to the value of this extra protection, but everyone agrees that colonies do benefit if they are protected from cold winter winds and will survive if there is sufficient honey and pollen. Experience over the past 100 years indicates that if bees have a good, soundly constructed hive; a healthy, young queen of superior stock; and adequate supplies of honey and pollen located in the proper area of the hive, the colony will survive a normal winter reasonably well and be productive the following spring.

The value of honey has increased to the point that commercial beekeepers and government researchers are reevaluating the overwintering situation. Considerable effort is being devoted to investigating the feasibility of overwintering colonies of bees in buildings or rooms with a controlled environment. Initial cost of facilities is quite high, since heating and cooling equipment is apparently required, but the savings in terms of honey production are believed to be sufficient to justify further studies.

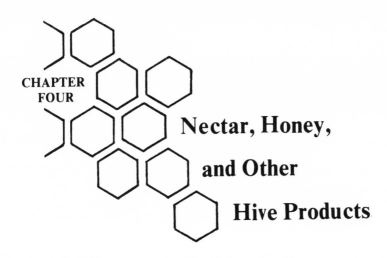

CHAPTER FOUR

Nectar, Honey, and Other Hive Products

Nectar Secretion

THE BASIS OF THE RELATIONSHIP between pollinating insects and plants is nectar. Nectar is an excellent source of energy, primarily carbohydrates consisting of sugars. While nectar contains other ingredients, such as traces of amino acids and minerals, their role is either minor or not understood. In their search for energy pollinating insects and other animals, such as moths and bats, accidentally transfer pollen from one plant to another. The tendency of foraging insects to concentrate on one species at a time is due to energy conservation. It is more efficient to work one species in bloom that contains nectar than to seek nectar-containing plants at random. This type of behavior enhances the possibility of cross-pollination. Some species are self-sterile; others are fruitful when self-pollinated, but the offspring are more vigorous when cross-pollinated. Cross-pollination, or cross-fertilization, is responsible for diversity in nature.

Nectar is secreted by special glands in the plant called nectarines. Some biologists believe that these glands export or excrete excess sugars. There are other types of export or excretory glands in plants. Plants growing in alkaline desert areas have glands that excrete excess salt. Glands in some plants are known to excrete or secrete wax, and some carnivorous plants secrete digestive enzymes that are used to degrade their victim so that they can absorb the nutrients. Sugars are manufactured during photosynthesis and transported in the sap to the growing points that would be the terminals of the plant and also of flowers. What is not utilized in growth is collected in the flowers as an excretory, or waste, product. Bees collect this waste product (nectar) and use it as a source of energy for themselves or for the colony. Extra nectar is converted to a concentrate and stored for future use. This stored plant-waste product is called honey.

Two types of nectarines are found on plants: floral, meaning "in the flower," and extrafloral, located somewhere outside the flower. Nectarines are associated with various parts of the flower and flowers are modified leaves, so it is understandable that through the evolutionary process some nectarines were located outside the flower as we know it today. Extrafloral nectarines are considered a more primitive structure, as they are not associated with pollination but function primarily as an excretory organ. Some extrafloral nectarines probably detract pollinating insects because they secrete nectar with a higher sugar content than that found in the flower. Under the microscope all nectarines appear similar. The nectar passes either through an opening or directly through the cuticle. The conductive tissue that feeds the nectarines varies from species to species. Physiological studies indicate that nectarines have a high rate of respiration, which suggests that nectar secretion is an active process. Nectar secretion is a complex process that is not fully understood. The quantity of nectar secreted appears to be directly associated with photosynthesis, sugar transport, respiration, and growth. Sugar appears to be manufactured in the leaves closest to the nectarine except in woody ornamentals and trees, in which it is derived from stored carbohydrates. In some situations the quantity of nectar is determined by the climatic and growing conditions of the previous year.

There are substantial differences in nectar secretion between lines or clones of plants. Hereditary factors are undoubtedly involved and nectar secretion varies among plant species, but it is related to flower age and to the shedding of pollen. Some plants secrete nectar for several days before pollination. The dandelion (*Taraxacum officinale*) secretes both pollen and nectar, yet depends on wind for pollination. Environmental factors, such as relative humidity, solar radiation, air temperature, soil water, soil temperature, and soil fertility are also related to nectar secretion. While no tests have been designed to study all these factors simultaneously, one simple fact seems clear: when nectar secretion is heavy, there are not enough bees to harvest it, and when it is poor, no amount of good colony management will correct the problem.

Can anything practical be done to improve or enhance nectar yield? Hereditary variations could be exploited. Current research in developing new varieties deals with such factors as protein content, total yield, disease and insect resistance, shipping and storage qualities, taste, and a host of other factors. Plants are not usually selected on the basis of nectar yield, although some work is now beginning to point in that direction. Research in soybeans (*Glycine max*) is directed toward developing hybrids, which would require the services of the indispensable honeybee. Some varieties are not visited by bees

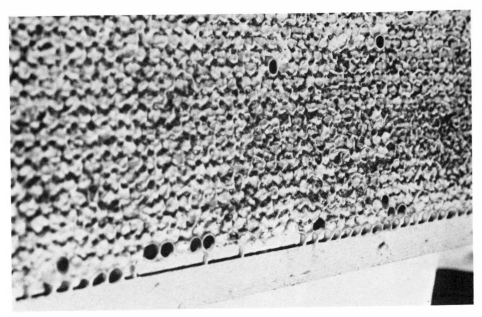

Sealed or capped honey. It is ready for extraction.

Uncapped honey. Bees will not seal or cap honey until it is dehydrated for storage.

due to their lack of nectar, so nectar secretion is an additional factor that plant breeders need to consider in selecting varieties. Honeybee activities and nectar secretion are influenced by the weather. Man still has not found an effective way to regulate or adjust the weather to suit his needs.

Nectar-producing Plants

Over 3,000 plant species are known to produce nectar, but 90% comes from about 100 species. Clovers contribute more to honey production in the United States than any other group of plants. Like the honeybee, clovers (*Trifolum* and *Melilotus*) came to the United States from Europe and found this continent quite suitable. While changing agriculture has forced a decline in clover-honey production in some areas, it is still one of the most important nectar-producing crops. White and yellow sweet clover (*Melilotus alba* and *officinalis*) are highly attractive to bees and on some occasions draw bees away from crops requiring pollination. The sweet-clover weevil has reduced interest in growing the crop as a soil-builder; the reduction of nitrogen-fertilizer costs also diminished the need for it, so bees have been deprived of this fine nectar plant. Economics prevents people from planting this or any other crop primarily for honey production, but there are acres of wasteland and land not suited for agriculture that do support fine stands of sweet clover. Additional acreages could be planted. The *Trifolium* clovers (red, white, alsike, Ladino, crimson, etc.) also produce large quantities of nectar. They are not as deeply rooted as the sweet clovers, and the nectar season is not as long. All clovers have a large number of blossoms per acre, so the crop can support a high population of bees. Clover nectar makes a white to light-amber honey with a mild flavor and usually brings a premium price on the open market.

Since 1950 alfalfa (*Medicago sativa*) has become increasingly important as a nectar-producing plant and in many areas is now the major nectar source. Alfalfa is a deep-rooted plant that can withstand drought better than many other plants. In livestock-producing areas where alfalfa is grown primarily as a forage crop early cutting reduces nectar yields. Alfalfa honey is light-colored, with a pleasing, mild aroma and flavor. Nectar is produced in large quantities and granulates slowly, so it makes excellent section-comb and chunk honey (honey marketed in the comb). As it is light in color and mild in flavor, honey packers like to blend alfalfa honey with the stronger-flavored dark honeys.

There are many species of asters (*Aster*) in the United States, most of which are highly attractive to bees, since they produce large quantities of nectar and pollen. Plants grow in wet to very sandy soil. Asters produce honey varying from very light and mild to very dark with distinctive flavor. Asters are relatively resistant to frost and in some areas will produce an abundant supply of nectar after goldenrods and other late-blooming plants have been killed.

Basswood (*Tilia*) produces a water-white honey with a very distinctive flavor. People used to mild honeys do not particularly like the biting flavor. The tree flowers in late spring or early summer and produces an abundant source of nectar when conditions are favorable. Basswood forests have all but disappeared in many areas, but recently some strains are being planted as shade and ornamental trees.

Citrus *(Citrus)* probably produces more nectar per acre than any other crop, even though it does not require pollination. (Some people feel that a high degree of cross-pollination produces seedy fruit). Most citrus honey is sold as orange-blossom honey, even though it may have come from limes, lemons, or grapefruit. Pure citrus honey is very light in color with a distinctive flavor. It can easily be discolored by other nectar if the flowering periods of other plants overlap.

Cotton (*Gossypium*) has been a major source of honey for many years. The honey is light amber in color with a mild flavor and compares favorably with the best grades available. Nectar secretion is variable, influenced by soil type, moisture, plant nutrition, and other factors. Cotton must be protected from a number of very destructive plant-feeding insects. Extensive bee kills have occurred, especially since farmers can no longer use DDT in their spray programs. Many replacements for DDT are highly toxic to bees.

Dandelion (*Taraxacum officinale*), while recognized as a pesty weed by many, is an extremely valuable plant to the beekeeper. It blooms early and provides an abundant source of much needed pollen and nectar. The honey varies in color from bright yellow to amber. Little is extracted, because most is used as food for developing larvae. Extracted dandelion honey has too strong an odor and flavor for bottling and is sold to bakeries, as it granulates quickly. The amount of dandelion honey and pollen harvested varies from year to year. In some parts of the country, dandelions bloom even when weather is variable. It is disheartening to see dandelions in full bloom, with the weather too cool for bees to forage.

Goldenrod (*Solidago*) is one of the more widely distributed plants in the United States. Different species vary widely in their value to beekeepers. Goldenrods in many areas flower late in the fall and produce nectar, which some beekeepers use for overwintering the colony; in other areas goldenrod yields no nectar or pollen. Goldenrod honey is usually quite thick and heavy, with a deep golden color. Some rate the quality poor; others prefer it to the light, mild clover honeys. It has a tendency to granulate quickly, sometimes before extraction. When this happens, the honey sometimes ferments in the comb. Much of it is sold to the bakery industry.

Sage (*Saliva*) is a California honey. While there are more than 500 species and most produce some honey, commercial qualities are only produced in California. Nectar production is dependent on rainfall. While the plant withstands drought, little or no nectar is produced unless 10" of rain occurs during the winter. The honey is white and does not granulate. Bottlers will pay a premium price for sage honey.

Soybean (*Glycine max*) acreage has increased during the last 20 years, and the demand for a high-quality plant protein will undoubtedly increase. Nectar production from soybeans is quite variable: some varieties produce little or none, while others produce a substantial amount. Soybean nectar produces a light-colored, mildly flavored honey.

Table 3 summarizes some additional plants and crops that are known to produce honey. While the color is easily determined, the flavor — mild, pleasant, bitter — is subjective and easily questioned. Some crops have been subjected to years of selection and breeding, so nectar and honey characteristics are quite variable, dependent on specific variety, and other factors.

Characteristics of Honey

Honey is the unique substance that first attracted man to bees. Throughout recorded history kings and philosophers have espoused the benefits, values, and virtues of honey: King Solomon's advice was to "Eat honey, my son, for it is good." Sacrifices made to gods in primitive societies included honey; Egyptian tombs contained honey and boxes of honey cakes; honey was served to guests at ancient Hindu weddings, and references made to it in the marriage ceremony; in nearby Bengal honey is a symbol of purity. The Greeks knew that honey relieved fatigue, and athletes drank a mixture of honey and water before major athletic events. Old age was generally attributed to fatigue, and the cure for this ailment was honey. At the time of Nero honey production was a viable industry: it was commonly bartered and exchanged in the marketplace. A number of northern European cultures recognized honey as the elixir of life. Malnutrition due to unbalanced diets and food shortages was widespread. Honey is assimilated into the body rapidly, rejuvenating a starved individual at least temporarily. Hebrew literature described the promised land as flowing with milk and honey. There is no mention of beekeeping, but there are several references to honey flowing from rocks. The Dead Sea scrolls refer to honey but not to beekeeping. Honey did not appear in China until about the 4th century A.D. It was probably imported from the West, as four distinct words of foreign origin are used to describe it. Mohammed claimed that honey is a remedy for every physical illness and the Koran for every illness of the mind.

The use of honey as an internal and external medicinal agent is old indeed. Its primary use was for stomach and intestinal disorders, followed closely as a treatment for respiratory problems. It was widely prescribed as a sedative as well as for disorders of the kidney and inflammation of the eye and throat. Its sweet, pleasant taste hindered its use as a vehicle for other medicines, because the prevailing idea in ancient times was that the harsher the taste, the better the medicine and the faster the cure. Hippocrates was an ardent advocate of honey, claiming that it creates heat, cleans sores and ulcers, softens hard ulcers of the lips, and heals carbuncles and running sores. It was also recommended as a cure for difficulty in breathing and to nourish and induce a good complexion. Dioscorides, a Greek physician of the 1st century A.D. who wrote *Materia Medica,* mentions honey, wax, propolis, and honey wine. Galen recommended a mixture of four parts honey to one part sea-tortoise gall to improve sight. He also claimed that honey and dead bees would grow hair on bald heads. Pliny mixed ashes of burned bees with honey as a cure for many ailments. The Arabs mixed honey with rose petals to cure tuberculosis. During the Middle Ages faith in the virtues of honey dramatically increased. It supposedly cured most ailments and was considered far better than wine, as it did not rise to the head. Among the ailments treated were snake bite, mushroom poisoning, and constipation. Honey was also used in surgery and burn therapy long before bacteria, molds, and fungi were known to exist. In the light of today's knowledge some of the claims for miraculous cures have some basis in fact. Because of its low moisture and low pH factor, honey inhibits the growth of microorganisms that are now known to cause problems in surgery and burn treatment. Its viscosity provides a good protective barrier over the wound and is easily removed with water.

In spite of its distinguished history, modern technology, and sophisticated laboratory equipment, what honey is and how it is made are not known. Laboratory analyses have identified 181 known substances in honey. Nectar is gathered by worker bees, mixed with enzymes from the head glands, deposited into hive cells, and stored for future use. Nectar contains a mixture of sugars, the specific kinds and ratios of which vary with the plant species. Sugar concentration ranges from 5% to 75%. In addition to the sugars nectar contains amino acids, enzymes, vitamins, minerals, and organic acids. In the hives enzymes split the sugars into glucose and fructose, and moisture is evaporated, hence a high-energy food is stored in its most concentrated form.

Honey is classified by the principal plant source from which the nectar is collected. While bees may work one plant species at a time, most honey originates from several plant types. Experienced individuals can readily identify the major honey types, such as clover, buckwheat, and citrus. Honey is also identified by the methods of production and preparation: extracted, spun, chunk, and comb are some popular types. The physical properties of honey can be easily measured with modern laboratory equipment. It is hydroscopic, as it will take moisture from the air, but the degree is dependent on two factors, the relative humidity and the concentration: for example, honey containing 17.4% moisture is in equilibrium with air at 58% relative humidity. The hydroscopic property of honey is a valuable asset compared to competing items, such as sugar and corn syrup. It has a tendency to keep baked goods and candies fresh and soft and is also used in some tobacco products.

Honey is a viscous product that can create problems in handling for the beekeeper as well as for the user. If honey is heated to 86°F., its viscosity is reduced so that it can be easily pumped, filtered, and bottled. Honey is heavy, with a specific gravity of 1.4 (1.4 times heavier than water). To some consumers the color of honey is important. In some cases the color-causing factor is known: for example, in buckwheat carotein imparts the color. In some types of honey polyphenols, iron, and tannins impart distinctive colors and flavors.

Honey contains at least 11 different mineral elements. Dark honey as a rule has a higher percentage of enzymes, dextrins, colloids, and other biologically active ingredients. The significance of some of these materials is controversial, and there are legal and moral restraints against making unsubstantiated claims. One biological agent that has been identified in honey is inhibine. Its antibacterial properties were discovered in 1937, and its effect was later discovered to be due to hydrogen peroxide, which is produced and accumulated in dilute honey by an enzyme. This enzyme is heat-sensitive and is removed in the pasteurization process. Honey is known to contain other materials exhibiting biological activity. Some of these activities have been described as root-promoting, estrogenic, cholinergic, and appetite-promoting. The significance of these activities is in some cases a matter of interpretation — some say imagination. Honey and nectar can also contain undesirable ingredients, some of which are highly poisonous to man and bees. Among the toxic chemicals are

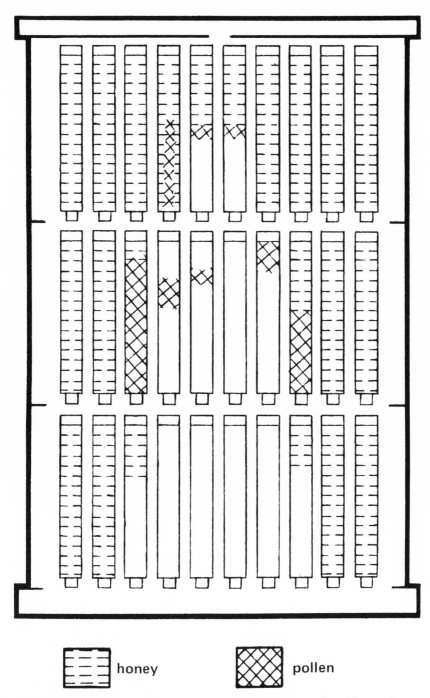

honey   pollen

End view of a hive in late fall. Honey is stored on the sides and top of the hive.

Electrically heated uncapping plane and knife for hand uncapping combs of honey.

Electrically driven uncapping knife.

Machine designed for automatically uncapping combs of honey. Radial extractor is in the background.

acetylandromedol, andromedol, anhydroandromedol, desacetyl, pieristoxin B scopolamine, gelsemine, and a glucoside arbeitin. Nectar from some Astragalus species and horse chestnuts (Buckeye) is toxic to bees.

Cane sugar has been known from antiquity and sometimes referred to as reed honey. It was probably introduced to southern China from India. At first cane was sucked or chewed; later sugar was extracted with water. The Greeks and Romans knew of sugar and referred to it as concrete honey. It was described as having the consistency of salt and it was said to be good for the belly and stomach. By the 14th and 15th centuries sugar commerce had increased, and it eventually began to replace honey as a sweetener. Beet sugar was a relative latecomer to the scene. The first commercial sugar-extraction factory was built in Selesia in 1801. Honey was probably the first animal product to be replaced by an exclusively plant-derived product. Other products that followed or will follow the same path are silk, wool, butter, and milk.

Handling Honey

The honey extractor was invented and patented in Europe in 1865. It permitted honey to be removed from combs without melting the wax. The extractor is basically a centrifuge. The wax caps of the cells are removed, and the combs put into a rotating machine. The honey is expelled from the cells by centrifugal force. Extractors of varying sizes are available, from the simple hand-operated type that holds two frames to the large, radial machines that are used commercially. The commercial beekeeper operator usually has special facilities for handling honey, while the hobbyist may satisfactorily use the home kitchen. The honey house has evolved over a period of years. It should have a bee-tight unloading dock or area to receive incoming supers, a heated holding room, an area for uncapping and extraction, and appropriate pumps and pipes to hold and transfer honey. Since honey is destined for human consumption, the facility must be designed, built, and operated in conformity with local, state, or federal health, sanitary, and safety regulations. While the hobbyist who extracts honey for himself and a few friends is not covered by such regulations, it is good business to handle honey properly. It would be tragic to ruin a wholesome food by improper handling techniques.

There are several ways to remove the wax caps from the combs. The hobbyist may use a large knife heated with hot water, or a steam- or electrically heated knife can be purchased. Commercial operators use a variety of mechanical and automated uncapping machines.

Capping wax is a valuable by-product, which should be kept separate from the wax recovered from broken combs and from scrapings from hives and frames. Wax derived from cappings usually brings a premium price. Honey can be separated from cappings by draining, straining, pressing, centrifugal force in an extractor, or melting under a closely controlled temperature of 125°F. (52°C.). Melted wax rises to the surface, since it is lighter than honey, and can be periodically drained off. At this temperature the honey will not be damaged.

Honey is a viscous product at low temperature. Many beekeepers hold supers in a room kept at about 90°F. (33°C.) before extracting. At this temperature the honey is easily and thoroughly removed from the comb. Some beekeepers force air through stacks of supers in the heated holding room to help reduce moisture. All honey must be strained after extraction to remove extraneous material. For the hobbyist a preliminary straining through a coarse strainer with 4 or 5 meshes per inch (1.5 to 2 meshes per cm) to remove large wax particles and a final straining through a screen or cloth with 86 to 100 meshes per inch (33 to 40 meshes per cm) is satisfactory. If the honey is heated to about 90°F. (33°C.), straining proceeds quite rapidly. The commercial processor uses a more elaborate system of screens, settling tanks, filters, and pumps. Honey is processed at a temperature of about 100°F. (38°C.) and, if it is to go into storage directly, put into a water-cooled heat exchanger for rapid cooling. It is handled in bulk, 55-gallon drums, and 60-pound (5-gallon) metal or plastic containers for the wholesale market. Processors that package products for the consumer market select the most suitable container sizes for their clientele.

The consumer has the right to expect a uniform-quality honey product. To assure quality, the Department of Agriculture has established honey grades and standards. Grading is voluntary and separate from identity, contaminant, and adulterant requirements. Some local governments and states have their own marketing regulations, which in most cases are similar to federal standards. Most states permit the sale of ungraded honey if it is so labeled. The Department of Agriculture classifies honey into seven categories or grades of honey, based on color: (1) water-white, (2) extrawhite, (3) white, (4) extralight amber, (5) light amber, (6) amber, and (7) dark amber. There are also four quality grades: (1) grade-A, or fancy; (2) grade-B, or choice; (3) grade-C, or standard; and (4) grade-D or substandard. Fancy honey must contain at least 81.4% soluble solids, have the flavor of the predominant floral source or a blend of sources, and be free from defects. Choice honey must also contain at least 81.4% soluble solids, have reasonably good flavor, and be reasonably free of defects. Standard must contain at least 80% soluble solids,

Table 3. Types and characteristics of honey.

PLANT	SCIENTIFIC NAME
apple - cherry - plum	*Malus*
asparagus	*Asparagus officinalis*
bird's-foot trefoil	*Lotus corniculatus*
blackberry	*Rubus*
cherry	*Malus*
cranberry	*Vaccinium*
cucumber	*Cucumis*
eucalyptus	*Eucalyptus*
gallberry	*Ilex glabra*
lima bean	*Phaseolus*
maple	*Acer*
melons	*Cucumis*
milkweed	*Asclepias syriaca*
mustard	*Sinapis*
palmetto	*Serenoa*
pine	*Pinus*
plum	*Malus*
rape	*Brassica rapa*
safflower	*Carthamus tinctorius*
smartweed	*Polygonum*
spanish needle	*Bidens*
sumac	*Rhus*
sunflower	*Helianthus annuus*
tamarisk	*Tamarix*
thistles	*Cirsium*
tupelo	*Nyssa*
vetch	*Vicia*
willow	*Salix*

COLOR	OTHER CHARACTERISTICS
light	fine flavor and aroma
light	very heavy
light	good quality
light	similar to clover
light	fine flavor and aroma
white	mild flavor
light amber	--
light to amber	very heavy
very light	pleasant taste
light to dark amber	mild flavor
pale amber	fine granulation
light amber	--
very light	slow to granulate
water white	granulates rapidly
white to light amber	mild and pleasing
dark	heavy, slow to granulate (honeydew)
light	fine flavor and aroma
water-white	aroma varies
dark	strong, unpleasant, disagreeable
light amber	unpleasant flavor
amber	strong, sometimes disagreeable
varies according to species	some good, others bitter
yellow	strong aroma
dark brown	minty aroma
light	good flavor, no aroma
light amber	does not granulate
dark	mild flavor
light	mild flavor

have fairly good flavor, and be free of major defects; substandard fails to meet any of these requirements. The honey grader samples each batch, matches the color with a reference standard, and classifies it into one of the standard colors. He then evaluates the honey on a scorecard system: the maximum number of points for each factor are: flavor, 50; absence of defects, 40; and clarity, 10. A score of 90 or higher rates a fancy grade; 80 to 89, choice; 70 to 79, standard; and below 70, substandard. Honey rated by an official USDA grader has two designations: color and grade. Standardized grading allows buyers and sellers to conduct business with confidence. It also allows beekeepers to interpret market prices and make decisions accordingly.

Standards of identity or composition are also used by regulatory agencies. In addition to protecting the public from harmful and unsanitary products they make specific products conform to recognizable and easily understood names. Honey is defined as a sweet substance, produced by bees from nectar of blossoms or from secretions of living plants, that is collected, transformed, and stored in combs, Specific details of moisture content and other factors vary from country to country: for example, in Canada honey cannot contain more than 20% moisture and 8% sucrose; in the United States 25% moisture and 8% sucrose are permitted. This moisture level would not meet USDA grade standards, but it can legally be called honey. In France it is legal to feed bees sugar, provided that the product is so labeled. Most other countries do not permit this practice. No country permits health or medicinal claims on the label of any honey product.

Most states have regulations that specify the kind of information required on labels of consumer products, such as contents; name and address of the producer, packer, or distributor; grade; and net weight. Federal regulations specify where and how the weight is to be expressed, as well as the size of the label and of the printing. While it is legal to sell ungraded honey, the label must specifically state that the product is ungraded.

Honey Types

Honey is a supersaturated solution of glucose. "Supersaturated" means that more material is dissolved than normally remains in solution. Such solutions are unstable, and excess material usually crystallizes or separates. Many honeys are in this category: some crystallize very rapidly, others slowly, still others not at all. The most favorable temperature for crystallization is 57°F. (14°C.). The size of the crystal varies with the conditions under which it was formed, and modern technology can control this to a degree. Fine-textured crystals are characteristic of unheated honey, or honey that is seeded either naturally or intentionally. This type is sold under trade names and sometimes referred to as spun or creamed honey. The retail market favors liquid honey, and processors try to package a product that does not crystallize. This is done by straining, heating, and filtration. Heating is done under closely controlled conditions so as not to affect flavor. A temperature of about 145°F. (63°C.) held for about 30 minutes dissolves whatever seed crystals may have been present. Some processors heat honey to 177°F. (77°C.) for 5 minutes and cool it rapidly to accomplish the same purpose. Care must be exercised so that the honey is not contaminated with microscopic crystals of honey, yeast spores, or even fine particles of dust. Any of these items can serve as seeds for recrystallization. Honey from various nectar sources differs in its rate of crystallization. No absolute figures can be given, because honey is seldom exclusively the product of single species of plant. If honey has been processed and allowed to recrystallize, the crystals are quite large and undesirable to most consumers. Honey can be reliquified by heating to 135°F. (57°C.). The shelf life of liquid honey (properly packaged) is relatively long but influenced by a number of factors. It darkens with age and with heat. Heating can also remove some of the delicate aromas. Honey can be kept in a freezer for long periods of time without crystallization or loss of flavor or aroma.

Fermentation and photosynthesis are the two fundamental chemical reactions in the living world. Photosynthesis creates food from inorganic materials, and fermentation breaks down food material into its original components. Photosynthesis uses chlorophyll and sunlight as energy; fermentation uses yeast cells and sugars. In fermentation yeast cells convert sugar to alcohol and carbon dioxide. In the presence of oxygen the alcohol changes to acetic acid (a component in vinegar) and, by further chemical degradation, to carbon dioxide and water (the building materials required for photosynthesis). The presence of acetic acid gives some fermented food a sour taste. Since honey is composed of sugars and yeast cells are practically everywhere, fermentation will proceed readily if proper temperature and moisture conditions are present. Honey processors guard against fermentation by pasteurization — heating until the yeast cells are destroyed — or by reducing the moisture content to 17% or less. Storing honey below 50°F. (10°C.) will prevent fermentation, but once honey is removed from this temperature, it begins, since yeast spores are not killed by low temperatures. Heating honey to 135°F. (58°C.) for 30 minutes or to 150F. (66°C.) for 1 minute kills yeast cells. It also dissolves microscopic crystals, so the honey processor performs two operations at the same time. This process produces a highly desirable liquid honey and assures a product that has a reasonable shelf life. Granulation sometimes increases the potential or risk of fermentation. As honey granulates, the moisture content of the nongranulated portion increases. If yeast spores are present and the honey is held at room temperature, fermentation will begin.

Chunk honey consists of a piece of honeycomb with capped honey that is cut to fit inside a jar. The container is then filled with pasteurized, extracted honey. The jar should contain at least 50% honeycomb by weight. In section-comb honey small section boxes are fitted into a specialized super. Bees build comb and fill it with honey. As soon as the honey is capped, it is removed. In producing bulk comb honey conventional frames are used. Once the honey is capped, it is sold in the frame or cut and packaged into plastic containers. A special foundation made of light wax is required to produce comb and chunk honey.

At the turn of the century most of the honey produced in the United States was in comb form. After World War I extracted honey became more popular for several reasons. Comb-honey production requires more labor and cannot be mechanized. It is more difficult to manage a colony producing comb honey, as it must be kept in a crowded condition that is conducive to swarming. The honey must be removed immediately after capping: if the capped comb is allowed to remain in the hive, the bees will stain the surface when they walk over it. Attention to such details is not feasible for producing a large volume of honey, so the producer must obtain a substantially higher price for the product, which many consumers are reluctant to pay, especially since they can buy extracted honey at a lower price. Another reason for the shift to extracted honey is that the public now has more confidence in the purity of honey. Food adulteration at the turn of the century was widespread, prompting Congress to pass the Food and Drug Act of 1906. At that time people had doubts about the purity of extracted honey and insisted on comb honey, since there was no way for an unscrupulous merchant to adulterate it. While there may be dishonest merchants and producers even today, government surveillance to assure compliance with the law gives consumers confidence that the product that they purchase is honey rather than a cheap substitute, and they tend to purchase the cheaper extracted honey.

Not all areas are suitable for producing comb honey. Regions with abundant and prolonged nectar flow, which produces honey that is slow to granulate, are best. Comb honey that has granulated in the comb has little consumer appeal. Areas with short or overlapping nectar flows of varying colors are similarly undesirable. Clover and citrus probably produce the best and most popular comb honey. Displays at county and state fairs, shopping centers, farmers' markets, and roadside stands sell substantial amounts of comb honey. Comb honey will always remain a gourmet item, one of nature's beautiful and wholesome foods. Honey in the comb tells its story better and far more effectively than the written or spoken word: the fragile, hexagonal cell of wax filled with fragrant, sparkling honey is still one of nature's true masterpieces of perfection.

Standards for comb and chunk honey comparable to those for extracted honey were developed by the Department of Agriculture. The top grade is fancy, followed by #1, #2, and unclassified. Color is classified into four categories: white, light amber, amber, and dark amber. As with extracted honey, each lot has a grade and a color. The exact description of each grade is quite specific, taking into consideration the condition and appearance of the cap, the cells, the honey, and the section box holding the comb. A detailed description of each grade is beyond the scope of this book. For further detilas contact the U.S.D.A. Agricultural Marketing Service in Washington, D.C. or your local land-grant-university or university-extension agent.

Selling Honey

The hobbyist with a few hives usually has no problem selling directly to friends, acquaintances, and neighbors once the word spreads that locally produced honey is available. It is estimated that 50% of the honey for home consumption is sold in this manner. The hobbyist can also sell to bottlers or packers, usually at wholesale prices in 60-pound containers or large drums. Buyers often want to see and test a sample of the product before purchasing. Bottlers often segregate honeys from different sources and demand a premium price for a specific floral source or color. Bottlers and packers sell honey under their own brand name or package it for specific customers. Gift shops and grocery chains usually sell products under their own company name. The gift-package honey market is large and highly profitable. This type of business can be developed in areas with a substantial tourist business. A number of local marketing cooperatives are engaged in selling honey. Members pool their honey, which is then sold under one brand name. One cooperative (Sioux Honey Association) operates on a national basis. About 25% of the honey market is sold through cooperatives.

Honey consumption in the United States is relatively low — less than 2 pounds per capita. Families that use honey use substantial quantities, yet most families use very little or none at all. Many youngsters reach school age without ever having tasted it. Beekeepers consider sugar as their chief competitor. The per-capita consumption of sugar has dramatically increased over the past 275 years: in 1700 the average consumption of sugar was 4 pounds per capita; by the Revolutionary War it had risen to 15 pounds; by the Civil War, to nearly 50 pounds. Today the per-capita consumption is about 135 pounds per person, about one-third of which is purchased directly as sugar. The remainder is used in food-manufacturing and-processing industries, such as cereal, soft drinks, baking, and canning.

Honey has several characteristics that make it an important ingredient in the baking industry. It is hydroscopic, retaining moisture and improving the browning qualities of the final product. Specialty products made of honey have appeared on the market from time to time. Many disappeared because of inadequate promotion or poor merchandising. Unless a product is patented, no company or individual is interested in promoting a specific product that a competitor could sell without the associated developmental and promotional costs. Honey has limited patentability. Some food manufacturers and processors are reluctant to use honey because of its viscosity and general difficulty in handling. To overcome this problem, a drying process was patented. Honey can be spray-dried: the final product is a flowing powdered mixture that is easily handled. Such developments undoubtedly expand the use of honey.

Many household foods routinely made with sugar or other sweeteners can also be made with honey. Your imagination is your only limitation. Spreads of butter, margarine, and peanut butter can be blended with honey; the percentage of each ingredient can be varied to suit the taste of your family. Flavors and spices such as cinnamon can be added to give it something extra. The shelf life of such spreads is limited, so they should be made in small batches and kept under refrigeration. Ice cream can be made with honey rather than suger, but refrigeration must be adjusted to a lower temperature. Since honey lowers the freezing point, the ice cream will be soft if conventional temperatures are used. This ice cream melts faster than the conventional product. Some cookbooks and recipes give directions on how to substitute honey for sugar. Honey contains 15% to 18% moisture, so adjustments can be made by changing the amounts of other liquids. The sweetening ability of honey is greater than that of sugar: a cup of honey weighs 12 ounces (and contains 9 ¼ ounces of sugar), while a cup of granular sugar weighs 7 ounces. Honey is acidic, and some recipes call for soda as a neutralizer: 1 cup of honey needs about 1/5 teaspoon of soda. Housewives tend to dislike working with honey because it is sticky. One handy hint is to measure fats or oils in the measuring cups before measuring the honey. It will then pour out to the last drop. Spoons should also be dipped in oil first.

Table 4. Physical and chemical properties of extracted (liquid) honey of average composition (based on 1 pound).

principal components		*percentage*
water (natural moisture)		17.2
levulose (d-fructose, fruit sugar)	38.19	
dextrose (d-glucose, grape sugar)	31.28	
sucrose (granulated, cane, or beet sugar)	1.31	
maltose and other reducing disaccharides	7.31	
higher sugars	1.50	
total sugars		79.59
acids (gluconic, citric, malic, succinic, formic, acetic, butyric, lactic, pyroglutamic, and amino acids)		0.57
proteins (nitrogen x 6.25)		0.26
ash (minerals -- potassium, sodium, calcium, magnesium, chlorides, sulfates, phosphates, silica, etc.)		0.17
		97.79
minor components		2.21
	TOTAL	100.00

pigments (carotene, chlorophyll and chlorophyll derivatives, xanthophyllis)
flavor and aroma substances (terpenes, aldehydes, alcohols, ester, etc.)
sugar alcohols (mannitol, dulcitol)
tannins
acetylcholine

enzymes: invertase (converts sucrose to dextrose and levulose),
diastase (converts starch to dextrins), catalase (decomposes hydrogen peroxide).
phosphatase (decomposes glycerophosphate), inhibine (antibacterial substance),

vitamins (thiamine, riboflavin, nicotinic acid, vitamin K, folic acid, biotin, and pyridoxine in small and variable amounts)

physical characteristics

caloric value: 1 pound = 1,380 calories
            100 grams = 303 calories
               1 tablespoon = 60 calories
1 gallon weighs 11 pounds, 13.2 ounces
1 pound has volume of 10.78 fluid ounces
1 gallon has equivalent of 11 pounds, 12 ounces granular sugar
Specific gravity at 20°C. = 1.41
pH = 4.5

Pollen

From the beekeeper's viewpoint pollen has one function: it serves as a source of proteins, fats, and minerals for bees. Pollen is rich in protein, but, like many of nature's products, it is not uniform in composition. Protein content can vary from 7% to 35%, but the average content is quite similar to that of beans, peas, and lentils. It is rich in lipids, minerals, and other elements. Since the composition of pollen varies considerably from species to species, a mixture of many types is necessary to give the colony a complete and balanced diet. Experiments have shown that 1 pound of pollen supplies about 4,500 bees. Based on these studies, a two-chamber colony uses about 44 pounds of pollen per year. The nutritional value of pollen has been evaluated through feeding trials. White rats appear to grow normally even if pollen is the only source of food, but male mice are retarded in growth and do not reproduce normally. Pollen has been reported to possess antimicrobial attributes, which probably affected the intestines of the mice. Opinions differ as to the nutritional value and benefits of bee-collected pollen in the human diet. While pollen has been used and recommended to cure and alleviate certain problems, its use is not based on scientific evidence, and some data points in the opposite direction. The walls of pollen grains are tough; they resist degradation in concentrated acids, hot alkali, and blender grinding. Pollen that has begun to ferment because of yeast cells is known to cause internal hemorrhaging. Humans thus do not readily· digest pollen grains, if consumed, and some may contain harmful substances.

Beeswax

Three kinds of waxes exist in nature: (1) mineral, derived from petroleum, commonly known as paraffin; (2) plant, one of which is commonly known as carnauba; and (3) animal — beeswax. Beeswax is a true secretion, produced by four glands on the bottom side of the abdomen of worker bees. Beeswax is a chemically complex mixture, consisting primarily of long-chain hydrocarbons, monohydric alcohols, organic and hydroxy acids, and diols. It has a specific gravity of 0.96 to 0.97 and a melting point of 147.9°F. (64°C.). (Impurities lower the melting point slightly.) Beeswax is soluble in chloroform, carbon disulfide, volatile oils, ether, and benzene but insoluble in water. Because of this characteristic it was and still is widely used as a waterproofing agent. Major uses of beeswax are in cosmetics, beekeeping (as foundation), and candle making. In addition to its symbolic use in church candles, it is an ingredient in ,high-quality decorative candles. It does not produce objectionable smoke and burns with an even, steady flame. Minor uses of beeswax are many and varied. Architects used to use it for models of structures, and sailing ships relied on it to waterproof and strengthen sails and lines. Beeswax is probably used in more different ways than any other single product. Among them are lotions, cold creams, ointments, salves, lipsticks, rouges, pill coatings, impressions and base plates in dentistry, castings in foundries, waterproofing, coatings for electrical apparatus, floor and furniture polishes, leather polishes, coatings for strings in stitchery and archery, art and craft items, adhesives, crayons, chewing gum, inks, basketball molding, grafting wax, ski wax, ironing wax, sealings for toilet bowls, and artificial flies.

Royal Jelly

Royal jelly, partly because of its name, has intrigued the uninformed and ignorant for years. It is secreted by glands in workers between 5 and 15 days old and fed to the queen throughout her larval and adult life and to larval workers for the first 2½ days. It is apparently synthesized during pollen digestion. It is moderately high in protein, creamy milky-white in color, and strongly acid, with a slightly pungent odor and a bitter taste. It is easily collected from larval-queen cells. Analysis shows it to contain 66% moisture, 12% protein, 5% lipids, 12.5% reducing substances, 0.8% ash, and 2.8% unidentified substances. It is rich in vitamin B and contains vitamins C and D but is lacking in vitamin E. It also contains an antibiotic substance (10-hydroxy-decanoic acid). Royal jelly is collected and used for queen-bee production and for research purposes. It is taken from a queen cell with a pipette or aspirator, filtered, and refrigerated or frozen for future use. Royal jelly has been promoted by unscrupulous individuals as a dietary supplement and as an additive to lotions, cosmetics, and creams. There is no evidence that it has any beneficial effects on organisms other than bees, and it is not recognized as a drug by the Food and Drug Administration or the American Medical Association.

Propolis

Propolis is a sticky, resinous material collected by bees from trees and other vegetation. It is used as a glue and caulking compound for sealing cracks and crevices within the hive. It is sticky when warm and makes it difficult to manipulate frames and hive bodies. This is a benefit in keeping hive bodies from sliding and slipping while colonies are moved. It has antimicrobial properties according to many controlled experiments. In Russia propolis is used in veterinary practice. It is added to ointments used for treating animal cuts, abscesses, and wounds, and doctors have experimented with an alcohol tincture for treating hearing defects and an anesthetic in dental practice. Stradivarius and other famous violin makers used propolis as an ingredient in violin varnish. Bees collect propolis from a variety of plants, depending on the needs of the colony and sometimes even if there is no apparent need. As with pollen, they carry it into the hive on their legs, but while a worker can unload pollen into a cell by herself, a worker bringing propolis needs help. Another worker takes hold of the propolis with her mandibles and tears it free from the collecting worker's leg. Propolis is not stored but used immediately. The worker removing it must find a place to use the material. It might be used to seal a crack between two hive bodies or to reduce the entrance to the colony or just stuck somewhere between the frames.

Special Beekeeping

Services and Operations

Pollination

THE AVERAGE BEEKEEPER is interested in producing honey, but the pollination performed by his bees is of far greater value to mankind. About 50 crops in the United States are dependent on or greatly improved by honeybee pollination. While other insects pollinate plants, the honeybee is one of very few that can be managed and synchronized with the development of crops. Renting colonies of bees for pollination began in 1910.

American agriculture has undergone a revolution in the last 100 years. Selection and breeding of high-yield plants, efficient use of agricultural chemicals, and a high degree of mechanization enable one farmer to feed approximately 50 people. This specialization has affected the beekeeper. Farmers growing crops that require pollination recognize the importance of bees and rent the services of beekeepers. (In other types of agriculture, such as corn, wheat, and barley, bees have no value). At one time bees gathered most of their nectar from wild or noncrop lands. This is no longer true. In some desert areas bees gather nectar almost exclusively from irrigated crop land. In other areas authorities estimate that only 50% of the nectar now comes from noncrop land. In the corn belt continuous corn on clean, cultivated fields and fields kept weed-free by herbicides has deprived bees of several excellent sources of nectar and pollen. Smartweed *(Polygonus)* used to be a major nectar producer in some areas in the late summer and early fall. To enable large machines to operate efficiently, many fence rows with brush, trees, and weeds were removed, depriving bees of wild foraging areas. Small farms consolidated into larger ones, also eliminating wild areas. This is especially detrimental if the farms plant large acreages of corn, wheat, rice, or barley. Changing crop patterns have greatly affected beekeeping and honey production. Buckwheat (*Fagopyrum esculentum*) was once a widely grown crop especially in the northern United States. It flourished on poor soil even if it was planted late in June or early July. The grain is high in fiber and relatively low in protein. Soil conditions can now be readily corrected, and corn and wheat are grown more profitably in many of these same areas. At one time millions of pounds of buckwheat honey were produced; now it is a rare item, and some people pay a premium price for this darkly colored, strongly flavored product. Buckwheat is still widely grown in Europe.

On a world-wide basis alfalfa (*Medicago sativa*) is probably the most valuable crop to mankind and as a forage crop for cattle because of its high protein. It is a legume capable of fixing nitrogen; it has a deep and dense root system, which is helpful in preventing erosion and in withstanding drought. The plant is somewhat resistant to insects and plant diseases. Varieties have been developed over the years that can be grown in practically every state and under a wide range of conditions. Unfortunately for the beekeeper the most desirable time to harvest alfalfa is before blooming or just as it begins to bloom. The later the harvest, the lower the protein content. Farmers who raise alfalfa for cattle feed are in competition with beekeepers who want to produce honey from the same fields.

Because alfalfa is such a widely grown plant, seed production is also big business, especially in the western states. Alfalfa not only produces excellent honey, but the plant must be cross-pollinated in order to produce seed. Alfalfa pollination presents some special and interesting problems for researchers. The alfalfa flower must be tripped and cross-pollinated. Tripping involves the release of the male and female parts of the flower, which are enclosed in a sheath or keel. They are kept under pressure and are nonfunctional until released. The sheath or keel is a specialized petal of the flower. The alkali bee (*Nomia melanderi*), the leaf-cutter bee (*Megachile rotundata*), and the honeybee can trip the alfalfa flower and cross-pollinate it. The alkali and leaf-cutter bees are very efficient pollinators of alfalfa, and methods of culturing both species are being researched in the western United States. To trip the flower, the bee has to enter it and press its head against the upper petals. The pressure trips the mechanism, opens the lower part of the flower, and releases the anthers (male part) and style (female part). When this happens, the anthers deposit pollen on the head of the bee. As she moves to other flowers, the process is repeated, and pollen is rubbed on the stigma (tip of the style) of the next plant, transferring it from one flower to another. Bees more than 3/8" long are generally more efficient trippers than smaller species, and bees less than 1/4" long do not trip at all. Bees do not like to do more work than is necessary to complete their mission of gathering nectar. This fact of life creates a problem for the alfalfa-seed grower. Older foraging bees learn to insert their tongue between the petals and to draw nectar without inserting their heads, so they trip only a low percentage of flowers. Younger pollen-gathering bees do enter the flower and are more efficient in tripping flowers, but the percentage of such bees is low. Alfalfa is not a highly attractive pollen plant, so workers do not hesitate to leave it even for obnoxious weeds such as thistles. In hot, dry weather alfalfa plants trip readily, and honeybees are effective pollinators under such conditions. Breeders hope to develop lines of bees that will be more efficient pollinators, and at the same time plant breeders hope to breed strains of alfalfa that can be readily tripped by the honeybee.

It is estimated that 38 million flowers must be tripped to produce 500 pounds of alfalfa seed per acre. Four to five strong colonies per acre can do this easily. The common practice is to place about ten colonies about 1/10 of a mile apart. Half are moved in when the first flowers show, and the remaining half about 10 days later. Movement of colonies must be synchronized with irrigation and insecticide treatments not only by the farmer renting the service but also with nearby neighbors. Alfalfa grown for seed can provide excellent honey, but the pollinator service is not necessarily able to capitalize on maximum production. For that reason the pollinator must charge a fee for his services.

Managing and handling bees for pollination is a specialized and mechanized business. Two to four hives are situated on a pallet, which is mechanically hoisted off the flat-bed truck-trailer with a boom-type fork lift and placed in a selected location within or along the edge of the field to be pollinated. For best results the colonies are moved from one location to the next at night. Rental fees vary from location to location. While pollination income helps stabilize the beekeeping enterprise, in most situations the beekeeper sacrifices some honey to provide the service.

There are inherent problems associated with pollination. As stated earlier, bees drastically reduce flight when it is too hot or too cold — 50°F. (10°C.) or 100°F. (38°C.). Winds above 15 m.p.h. slow bee activities, and above 25 m.p.h. they practically stop flying. In good weather pollination can be completed very quickly. The pollen can be affected by too much or too little moisture, and frost can kill fruit bloom. Strong colonies are essential to good pollination, which is dependent on the number of field bees collecting pollen and nectar. If the colony is small or weak, most bees stay inside the hive and tend brood. It is also important to have bees of proper age in the pollinating colony: if all bees were newly emerged, they would be less likely to forage. The time in which colonies are moved to a crop can be very important, especially if there are competing plants in the immediate area. If they are moved too far in advance of flowering time, the bees might start working other plants and neglect the principal crop. A rule of thumb is to place 25% more bees on the less attractive crop. Pears are one example: the nectar contains a low percentage of sugar, and bees prefer apples or other crops that flower at the same time. If none are available, then they work pears. Highly attractive crops such as cucumbers, which have a relatively small number of flowers per acre, can be adequately pollinated with one colony to every 3 or 4 acres, while pears may require two to three colonies per acre.

Commercial Queen Rearing

Raising queens is big business, especially in the southern United States. It is a highly developed art, and some operations are equipped to produce large numbers of queens on a time schedule. This section is not intended to fully describe commercial queen-rearing operations but to give an overview of the principles involved. A hobbyist who wants to rear a few extra queens might utilize the basic ideas, but his operation would be different. As with any plant or animal breeding or selection process, only the best should be used as the parent or breeding stock for future generations. The commercial queen producer can purchase breeder queens developed by the United States Department of Agriculture. Some USDA parent-or breeder-queen lines have undergone years of selection based on actual production. These breeder queens are used to head colonies, and the offspring from these colonies should be used to produce queens. Commercial queen production is concentrated in the south primarily because of climate. The queen must fly out to mate, and in northern areas the weather during the critical time is unpredictable. Queen- and package-bee production does provide off-season employment for some beekeepers. This is especially important for beekeepers who wish to be fully employed all year.

The commercial queen producer needs special equipment or standardized equipment modified to suit his needs. The breeder queen is confined to a single dark comb with queen-excluder material. Dark

Table 5. Protein and total digestable nutrients of alfalfa at various stages of growth.

	% Protein	% TDN
vegetative	31.2	87.0
prebud	26.8	82.6
early bud	22.2	72.8
first flower	18.2	60.1
full bloom	15.5	60.0
green seed pod	14.8	57.4

Table 6. Per-colony honey-production and labor-input (time) averages.

	two-queen		single-queen		package colony	
	pounds honey	minimum	pounds honey	minimum	pounds honey	minimum
1967	292	48	237	35	126	13
1968	266	46	169	33	106	12
1971	295	43	160	27	127	12
1972	246	40	82	26	-	-
1973	317	45	204	25	163	14
1974	224	35	138	29	75	9
average	280	43	168	29	55	12

comb is used so that eggs are readily seen. On the second day this comb is replaced by another comb, and the comb with freshly laid eggs is moved to another part of the hive. Each frame must be marked or identified, and records maintained. On the third day the process is repeated. Eggs in the comb laid on the first day begin to hatch, and when the larvae are 12 hours old they are ready for transfer, or grafting. By confining the queen to one frame per day the exact age of the larvae is known. Queen producers can make their own cups (cells) or purchase them from suppliers. A small quantity of royal jelly, slightly diluted in water, is placed in each cup cell. With a grafting needle young larvae about 12 hours old are removed from their normal brood comb and transferred to queen cup cells. It is important that the larvae be between 12 and 18 hours old, never older than 24 hours. Larvae in cup cells are now placed in starter hives. Bees in the starter hive continue to feed the larvae royal jelly and begin raising them as their own queen. It is important that the starter hives have a good supply of young bees with sufficient pollen and that they are kept queenless. It is also important that the starter hive not be overworked — too many grafted larvae should not be put into a starter hive. There are no hard-and-fast rules: it is up to the queen producer to make sure that the starter hive has a constant supply of young bees. This can easily be done by regularly adding combs of brood with emerging bees. The queen cup cells are kept in the starter colony for 24 hours, then transferred to the builder colony. Its condition is also very important. It must have a constant supply of young bees, pollen, and honey. Queen producers have developed a system of placing and rotating their queen cup cells, so they know the exact date on which each operation is performed. On the tenth day the sealed queen cells are transferred to nuclei for emergence. Each queen cell must have its own emergence or mating nucleus. Some queen producers use a double-graft system to produce queens. The larvae are transferred to a fresh queen cup cell, with fresh royal jelly, 24 hours after the first graft, then put into the builder colony to complete their development. Double grafting, while it involves more labor, produces superior-quality queens.

While the basic genetics of a queen cannot be altered by nutrition, there is ample data to indicate that a queen with excellent potential can be reared under poor conditions and become a poor queen. Developing queens need an abundant supply of royal jelly during development. The ovaries of a queen bee have the potential to develop 200 egg tubules. Queens reared under poor conditions develop only about 125 tubules, whereas the average queen, reared under good to excellent conditions, will have an average of 180 tubules. Queens reared under poor conditions are often replaced by workers much sooner than those raised under good conditions.

As emergence time approaches, the capped queen cells are transferred to the mating yard. Nuclei can be made to use standardized combs with conventional hive bodies, which are modified to have three or four separate chambers, each with its own entrance. Each chamber must be stocked with a population of bees and with a queen cell about to emerge. Between 4 and 7 days after emergence the new queen flies out to mate and returns to her hive to lay eggs. A mating yard may have up to 2,000 mating nuclei, so it should be stocked with one or several colonies with a high number of drones. Controlled mating with specific types of drones is possible but only in isolated areas. Two weeks after the queen cells are placed in nuclei the queen should be mated, laying, and ready to be caged for shipping. In this operation the queen producer's reputation is at stake. Only good, well-formed queens should be shipped, and only an experienced and properly trained person can make this judgment.

There are many variations in the methods of raising queens. Authoritative articles and books on the subject are available for those who wish to consider queen rearing as a business venture. The hobbyist who purchases a few queens each spring should understand at least what is involved in rearing queens and that there are excellent, poor, and careless producers.

It is simple to rear a few queens for your own use. Select only the best colony as the parent. Remove the old queen and two or three frames of brood with attached bees and place them in an empty brood chamber (the further away from the hive, the better). The bees and the old queen should have honey, pollen, and a separate flight entrance. Some will make small hive bodies, called nuclei boxes or chambers, for rearing queens. The queenless colony will now produce 10 to 12 queens. After the queen cells are sealed, they should be carefully removed from the comb. Save only large, well-formed cells and destroy the rest. Attach one queen cell on a separate comb by removing a section of cells measuring about 3/4" x 1 1/2" and attaching the queen cell in this area. Place this comb into a queenless colony or into a special nucleus with some workers, pollen, and honey. After the new queen emerges, the queenless workers will accept her as their own queen. She will fly out, mate, return, and begin to lay eggs. Each queen cell should be attached to separate comb and placed in a separate nucleus. The developing queen inside a queen cell is quite delicate and should be handled with care. If the weather is cool, be sure that cells do not chill. As soon as you remove the queen cells from your breeder colony, you can return the old queen, with attached bees and brood, to her former colony and repeat the process in a few days. The time required for the queen to develop is predictable. The cell will be capped 8 days after the egg is laid and emerge on the 16th day. By removing the old queen on a specific date you know the exact date that the cell will be sealed and the day that the queen will emerge. You can actually schedule your cell-transfer date and know when to expect the queen to emerge and fly out for mating. It is important to develop queens in a large or strong colony. Extensive tests prove that queens developed in small or weak colonies are smaller and lay fewer eggs during their lifetime.

Bees have been transported all over the world. Early travelers undoubtedly took complete hives, boxed or crated in a suitable fashion for travel by ship, boat, or caravan. Shipping bees in simple wire-screened cages began in 1879. As with any new development, technical difficulties had to be overcome. In 1912 the first successful large-scale shipment of bees was made from the southern United States to points further north. The demand for packaged bees grew steadily, peaking in 1947. At one time, 80% to 90% of the packages were shipped by train; today trucks and air transportation have become increasingly important. Packages are now standardized (10" x 5 1/2" x 14") and can hold a 2-pound tin can with sugar syrup for food in transit. There are a number of ways to fill packages, and efficiency largely depends on the imagination of the beekeeper involved.

Preparing colonies for package production begins in August or September. Poor queens should be replaced, and hives checked for adequate pollen and honey. About 10 weeks before taking bees, the colonies should be stimulated to rear brood by supplementary feeding of pollen supplements or by placing combs containing pollen near the brood nest. About 4 pounds of bees can be removed at 10-day intervals from a good colony. The average commercial package-bee producer sells about 10 pounds of bees per colony. The maximum taken from one colony was 37 pounds, averaging 32 pounds over a 60-day period. After crating the bees each package is supplied with a caged queen. It hangs in a slot cut in the top of the package next to the feeder can. After the queen is in place, the feeder can is set in the package and the cover nailed — the package is ready for shipment. Beekeepers purchasing 200 or more packages often find it economical to pick up their own bees. Many combine the package pickup with a spring-vacation trip to the south. It is important, however, to bring the bees to their destination with minimal stress. If the weather is cold, they should be protected; if it is hot, they need ventilation during the ride north.

Do northern beekeepers have to haul bees from the south? There is evidence showing that beekeepers who overwinter bees in northern climates can produce some surplus bees with proper early-spring management. The colony must be inspected in March, then fed either pollen, pollen supplement, or pollen substitutes. Early-spring feeding stimulates brood rearing, and by mid- to late April colonies will be strong enough so that 2 to 4 pounds of bees can be removed without damaging their productive potential later in the season. It is important that northern producers make provisions to furnish queens with their packages. Queen production is impossible in northern climates at that time of the year because the weather is too unpredictable for mating, so beekeepers still have to depend on southern producers for this specialized service.

The purpose of instrument insemination is controlled mating. In nature the queen mates by flying away from the colony. Drones are attracted to a queen from great distances. Because of this behavior it is impossible to control breeding, as the origin of drones is unknown. Attempts to inseminate queens with instruments were made over 180 years ago. It was not until 1927 and 1928 that significant breaththroughs were achieved. More was known about anatomy of the reproductive organs and better instruments were developed, making a workable technique possible. Progress was slow, primarily because basic information on the physiology of the bee and its semen was lacking.

The technique is basically quite simple. Semen is collected from a drone in a small syringe and placed in the oviduct, from where it passes into the spermatheca of the queen. The spermatheca is a unique structure. It is designed to hold and store sperm for the entire life of the queen. Sperm cells are produced in testis and pass into the seminal vesicle, where they accumulate and mature. It is important to use drones at least 12 days old. Sperm are collected with a small syringe by stimulating the drone by decapitation, pressure on the abdomen, chloroform fumes, or electric shock. The queen to be inseminated must be of proper age. Best results have been obtained after 7 to 10 days, but early or late inseminations have also been successful. The queen is placed in a small tube and anesthetized with carbon dioxide. Using a set of specially designed instruments and working under a binocular microscope, the sting chamber is opened and kept spread. The syringe tip covered with semen is inserted into the oviduct, and the semen deposited. The volume of semen used depends upon the inseminator. In bee-breeding work and genetic studies it is often necessary to inseminate a queen with semen from a single drone. In such cases only a small volume of semen is available. Under natural conditions the average queen receives about 5 million sperms in her spermatheca, but with instrument insemination, using 4 microliters of semen from three or four drones, the sperm count is about 3 million. Queens have been known to fly out and mate after instrument insemination, but they will not mate naturally if they are inseminated with 8 microliters of semen. It is estimated that between 3 to 5.4 million sperm will reach the spermatheca with an 8-microliter solution.

Instrument insemination has been perfected to such a degree that technicians can inseminate queens and obtain a very high percentage of fertilized queens. And the queens are as productive as those naturally mated. There are differences of opinion as to the value of instrument-inseminated queens in honey production. Some feel that its value lies primarily in producing breeder queens with known parentage and in genetic studies. Once a breeder queen is mated with drones of known ancestry, queens produced from her colony (daughters) can be mated naturally. They retain enough desirable genetic characteristics to make the purchase of queens from these breeders worthwhile.

It is normal to have one queen in a honeybee colony, but a mother-daughter-queen combination laying harmoniously in a single colony is not unusual. Experienced beekeepers frequently observe this phenomenon, which is the basis of the two-queen system. The idea of using two queens in one colony was initially studied in 1890 and again in the 1930s and 1970s. Labor input and production were documented in the latest study.

Operating and maintaining a two-queen colony is basically quite simple if the principles of bee management are understood. The time to start the colony is about 2 months before the main nectar flow. The colony is divided in the spring, as previously described, by using an inner cover with the hole in the center closed. The old queen is placed in the lower chamber. Both the upper and the lower chamber should have pollen, honey, and a separate entrance. The second queen is introduced into the upper queenless chamber by the slow-release method. After the second queen is released and laying — in 2 or 3 weeks — the inner cover is removed and replaced with a queen excluder. Each queen will remain in her own area of the colony and continue to lay, and workers can move through the queen excluder and the entrances easily. Each section must be handled independently in terms of colony manipulation. If the top queen is too high, the top brood chamber should be reversed independently of the lower brood chamber. The lower brood chamber must also be inspected and reversed when needed. This is the part of the two-queen management system that some beekeepers dislike. It is harder to inspect and reverse a two-queen colony than a single-queen colony. A two-queen colony can swarm as readily as a single-queen colony.

About a month before the honey flow ends the two-queen colony is no longer advantageous, and most beekeepers revert to single-queen status. The advantage of the two-queen colony is lost since it takes about 5 weeks to develop a foraging bee (3 weeks from egg to emergence and 2 weeks from emergence to field work). At this time either the queen excluder is removed or the second queen is taken out of the colony. The colony is now a single-queen colony with an unusually large number of workers. Honey production is directly related to the number of workers. A two-queen colony needs less capital equipment (one less top cover and bottom board per colony), stores more pollen for the following year, and produces more honey. The colony must be placed on a firm foundation or it will tip over. Some feel that a bench or some other object to stand on when adding supers is desirable as the colony will be quite tall. The two-queen-colony system illustrates how a wild animal (the honeybee) can be successfully managed to benefit mankind. Beekeepers who have never tried the two-queen system can be assured that it offers a unique experience.

CHAPTER SIX

The Enemies of Bees

LIKE ALL LIVING THINGS, bees have their share of enemies. Am ng their parasites and predators are viruses, fungi, bacteria, protozoa, other insects, some birds, a few mammals, and man. Man, through careless management and ignorance, is by far the most destructive creature that bees encounter. One solution to this problem is for knowled/eable beekeepers to educate their less enlightened or less fortunate colleagues. Within the last 50 years real and imaginary conflicts of interest have developed between beekeepers, other businesses, leisure-time activities, and agriculture.

Mammals

Among the mammals the skunk rates as the worst enemy of bees. This is partly due to the fai that the skunk is one of the more widely distributed mammals. Some states protect skunks by law. Many states have provisions that allow beekeepers or other individuals to protect their property if they are destructive in any way. If a beekeeper has difficulties with a skunk (or skunks), the first recommendation would be to contact the agency responsible for protecting skunks. Skunks are by nature insect eaters. Once they are educated in the technique of capturing and eating bees, they can drastically reduce the population of a colony. Skunks scratch on the front entrance of the hive, and, as the bees come out to investigate the disturbance, they are captured and eaten. Some beekeepers believe that skunks agitate and irritate the colony so that when the beekeeper makes his routine inspection, the colony is unusually defensive and difficult to manage. Hives that are being visited by skunks usually have scratch marks made by claws around the front entrance. Skunks can be fenced away from individual hives and even yards. It is probably cheaper to eliminate them by trapping or poison if allowed in your area.

Bears can be extremely destructive to a hive or bee yard. They are omnivorous and frequently dig into rotting stumps and logs for a variety of insects. It would be more productive and probably easier to dig into a hive of bees. Some scientists believe that the European bear evolved as a true predator of bees. Bears are known to seek out a wild colony in some very unusual places. The American bear evolved differently, however, because the honeybee did not arrive on the North American continent until about 1620. Nevertheless, the American bear likes honey and/or brood. It is known to search out or smell a colony from a long distance. Once in a bee yard, a bear can raise havoc. Bears primarily eat brood comb, but in their search they usually destroy the colony and ruin the combs and frames. One effective means of protecting a yard from bears is an electric fence. Multiple wires (or better, a wire mesh similar to poultry netting) are erected around the entire yard. Bears' feet apparently have a moderate amount of insulation, especially on dry ground. To make the electric fence effective, a wire netting or mesh has to be laid around the outside perimeter as a ground wire in such a fashion that the bear stands on the ground wire when it touches the electrified fence. If this is not done, most electric fences will not slow the bear's approach to the colony. Brood comb and honey have been used as bait for bears, especially by bow-and-arrow hunters. This practice has been frowned upon by some sportsmen, and a number of states have outlawed the practice. While the technique is effective, there is an inherent danger of spreading bee diseases.

Mice can be very destructive to a honeybee colony. While they do not eat or kill bees directly, they enter the hive in late fall or early winter to build a nest in the corner of the brood chamber. During warm weather bees can protect the colony from intruders, but this unwanted guest enters when the bees are clustered. Mice enter beehives only because they offer a sheltered cavity for nesting. During the process of building a nest substantial brood comb is destroyed. While bees will reconstruct the damaged or destroyed comb, the new construction is often very uneven and usually contains drone cells. Mice undoubtedly enjoy some honey and pollen and consume some of the dead bees on the bottom board, but their primary interest seems to be shelter. With the approach of cold weather, some beekeepers in northern areas reduce the bottom entrance to the hive with blocks of wood, allowing an opening only about 3/8" to 1/2" long and 2" to 3" wide. Others place hardware cloth, with a 3/8" to 1/2" opening, in front of the entrance as an effective barrier against mice.

Birds, wasps, hornets, robber flies, dragonflies, toads, spiders, and a host of other organisms that feed on insects nonselectively can be potential enemies of bees. While the number of bees that any one of these creatures can capture may appear insignificant, consequences to the colony could be more serious if one of the bees happened to be a queen on her mating flight. Microorganisms that affect bees are usually more costly to the beekeeping industry than are the large predators.

Nosema

Nosema was mentioned earlier in the context of colony management. The disease is not often recognized, because the symptoms are such that it rarely kills the colony outright. The colony becomes weak, and some beekeepers attribute this condition to overwintering stress, dysentery, spring decline, queenlessness, and many other reasons. The casual organism is a protozoa (class Sporozoa), a single-celled animal related to the malarial parasite of humans as well as to the amoeba and paramecium usually studied in elementary biology. The nosema organism was studied in 1909. The species (*Nosema apis*) is harmful only to the honeybee. It is found in all parts of the world where honeybees are kept. There are other nosema organisms present in other insects, but they do not affect honeybees. The parasite lives in the lining of the midgut in the digestive tract. It has two stages in its life cycle, vegetative and spore. Bees must swallow spores to become infected. Once the spores germinate, they penetrate the cells lining the midgut. After a period of growth, usually 6 to 10 days, the cells are filled with many more spores. The cells break and spores are spilled into the intestine, where they germinate and reinfect other cells in the same bee or are voided in the feces. If this is done inside the hive in winter, other bees will become infected. The cells lining the intestine are destroyed, so the bees cannot digest or absorb all the required food nutrients. Nosema creates a type of dysenteric condition. The fact that the bee cannot digest and absorb all its nutrients, plus the dysentery, shortens its life by as much as 50%. Should the queen bee become infected, she will be replaced in 2 to 7 weeks. If this happens in midwinter or very early spring, the colony will become queenless and usually die. Since infected workers have only about 50% of their normal life span, the colony population dwindles even though there is still some brood rearing. In late winter or very early spring it is quite easy

Table 7. Brood diseases.

characteristics	American foulbrood	European foulbrood
appearance of comb	capped brood is discolored, sunken or punctured cappings	primarily uncapped but some capped in advanced cases, cappings sunken or punctured
age of dead brood	old capped larvae or young pupae	young larvae, occasionally older
color of dead brood	light brown, coffee-brown, to almost black	yellowish white to brown, almost black
consistency of dead brood	soft, becoming sticky to ropy	watery to pasty, not ropy
odor	dead animal tissue	sour-fermented
scale	uniform, flat, on lower side of cell, adheres tightly to cell wall	twisted in cell, easily dislodged from cells, sometimes rubbery
control	burn if required by local laws	burn if required by local laws
treatment	sodium sulfa thiazole[1] oxytetracycline[2]	oxytetracycline[2]

[1]sodium sulfathiazole is a stable material that can be mixed in sugar syrup and fed to bees. Use ½ teaspoon per gallon of feed: higher dosages are no more effective and exhibit some toxicity. It can also be mixed with equal parts of powdered sugar, and about ½ teaspoon of the mixture dusted over the top bars of the hive. The dusting treatment is used if bees require no supplementary feeding. The treatment should be repeated three to four times at 4- to 5-day intervals.

sackbrood	chalkbrood
capped brood, scattered cells with cappings punctured with two holes	scattered cells often on perimeter of cluster
old uncapped larvae	young pupae, old uncapped larvae
grayish to straw-colored, brown to black	chalky white to gray, usually fluffy filament on surface
watery and granular, tough skin forms sack	firm pellet
none	none
easily dislodged, head curled upward, rough texture, brittle	easily dislodged, usually seen on bottom board or at front entrance
none available	none available

[2]Oxytetracycline (Terramycin) is used at a rate of ½ gram *active ingredient* per hive. The product is available in two formulations (animal-grade), TM-10 (TAF) or TM-25 containing 10 and 25 grams, respectively, per pound. One pound of TM-10 can be diluted with 3 pounds powdered sugar, or ½ pound of TM-25 mixed with four pounds of powdered sugar. About 3 tablespoons of the sugar-drug mixture can be dusted over the top bars of the hive. This treatment should be repeated about three times. Suppliers and regulatory agencies may have slightly different procedures for using medications.

to diagnose the disease. The combs are often soiled because workers cannot fly out to void feces. If weather permits flight, the entrance and the front of the hive are often covered with excreta. The disease organism spreads within the hive through combs soiled by the infected workers. The young workers usually clean and polish the comb and become infected in the line of duty.

In areas where bees can fly out all year the symptoms of nosema are somewhat different. Sick bees can usually be found crawling around in front of the hive. Some attempt to fly, and the entire body trembles. Further north these same symptoms can be noted prior to nectar flow. The disease incidence within the colony tends to decrease during the summer. The disease can be diagnosed in the field quite accurately. The thorax of a suspected bee is held, and the abdomen is slowly pulled away. The intestinal tract of a healthy bee will be firm and have several constrictions. The midpart of the intestine of a bee infected with nosema will be swollen, and the constrictions will be missing. If the midgut is crushed on a glass slide in a drop of water and examined under a microscope, the spores will be clearly visible. A magnification of about 400x is needed to positively identify the spores, but experienced individuals can use a lesser magnification (40x) and still be certain of their findings. The nosema spores are much larger than other microorganisms encountered under similar conditions.

The disease can be arrested but not eliminated from a colony with a drug called fumigillian. The drug has no effect on nosema spores but does protect bees from the vegetative stage. The drug must be provided continuously — for example, mixed in sugar syrup. Fumigillian can also be mixed into pollen substitutes or supplements. The only problem with this technique is that a colony weakened by nosema tends to eat less and continues to decline. Management can play an important part in keeping the incidence of the disease low. Nosema is basically reduced by using good management procedures: (1) keeping and maintaining only strong colonies with a prolific queen and (2) protecting the hive from cold winds. The location should have good air drainage and be sufficiently open so that bees can fly out on sunny days. The greater the number of infected bees that can leave the hive, the less infection will remain in the colony.

Fumigillian is fed to the colony late in fall and again in spring in a sugar-syrup mixture. In the fall treatment the syrup-fumigillian mixture is stored in and around the brood nest. It is the material that will be used as food during the winter. Detailed studies and experiments over many years leave no doubt that the drug treatment significantly reduces nosema infection in the colony. Most colonies survive the winter in a much stronger condition. Early-spring treatment with fumigillian is highly beneficial to the colony. At this time the colony is rapidly expanding, and young workers are cleaning and polishing combs. Some of these combs will undoubtedly contain nosema spores. A syrup-fumigillian mixture given at this time of the year will protect the young workers during this critical time. In northern areas beekeepers who rely on package bees shipped from the south each year experience great losses from nosema. A few bees with nosema can infect the entire package unless special precautions are taken. Many beekeepers order 10% to 20% more packages than they ordinarily need, assuming they will lose some to nosema. Purchases of queens also suffer substantial nosema losses. The new queens somehow contact nosema and are replaced in two to seven weeks after installation. Packaged-bee and queen producers could probably obtain a premium price for their products if they could assure their northern beekeeping customers that their bees were protected with fumigillian.

Bacterial Diseases

The beekeeping industry has changed dramatically in the last 30 years. Bee breeders are constantly looking for new lines (genes). This involves shipping breeding stock to great distances. Economical, rapid transportation enables beekeepers to move large numbers of colonies hundreds of miles in a short time. These practices create new opportunities to spread infectious diseases unknowingly or unsuspectingly. The same is true of shipping used equipment and trapped pollen. For these reasons the beekeeper, whether he is an amateur, hobbyist, or commercial businessman, should always be on the lookout for problems or unusual conditions in his colonies. A beekeeper cannot rely on a professional such as a veterinarian to diagnose disease problems but must be his own field diagnostician and, like a professional, willing to submit samples of material to laboratories for confirmation. Through the land-grant-university system expert service and advice is readily available. Your local county-extension agent or farm advisor can usually put you in direct contact with a service in your state. On-the-premises inspection is available in many states. The cost of the inspection is usually paid for by the beekeeper, although this procedure varies widely from state to state and even within one state.

The present method of detecting diseases is slow and laborious. It amounts to visually examining each brood comb in a colony. The suspected combs are sent to a laboratory for microscopic examination. Needless to say, a simple, rapid screening test would be extremely valuable. Prior to the discovery of antibiotics a beekeeper who had infected colonies was required to destroy the colony by burning it under supervision, and burying the ashes. Some states paid a modest indemnity to the owner of destroyed hives. The law authorizing search-and-destroy permitted an inspector to enter the apiary. If infected colonies were found, the yard was placed under quarantine. No bees or equipment could be moved; infected colonies were destroyed, and disease eradicated. Some states keep records of disease reports by registering yard locations. A purchaser of bees or equipment can determine the past history of each yard. The procedure for destroying infected colonies was to kill the bees with cyanide gas, dig a pit at least 18" deep, and feed the diseased combs to a fire. The hive bodies, bottom board, and top covers could be scorched with fire; the ashes were then buried. If bees were gassed at night or on a day of no flight, the eradication was effective. If eradication was attempted on a day when field bees were working, results were not always satisfactory. Bees returning to the colony and finding their own hive missing eventually joined other colonies. If they were carrying disease organisms, they could infect other adjoining colonies.

As with many laws, ordinances, and rules, the degree of enforcement varied from community to community. Needless to say, the bee inspector was not the most popular person in town. Duties were sometimes performed under the protection of the sheriff. Some states have modified their laws pertaining to bee inspection to be consistent with developing technology; others have not. Some basic laws were written years ago in such a manner as to allow the introduction of new techniques and technology without having to go through the legislative process. The basic bee-inspection law was undoubtedly justified. Diseases affecting larval bees at one time nearly eliminated the beekeeping industry. The turnaround was gradual. In some states diseased colonies now comprise less than 1% of inspected colonies, including migratory colonies that move from one state to another. Most states require a point-of-origin inspection. Yards suspected of having a disease problem (by owner request or by complaint) are inspected, and the inspection reports are used to determine the severity of the disease.

Diseases contracted during the larval or pupal stages are called brood diseases. They may be caused by bacteria, viruses, or fungi. Two of these diseases are so important that every beekeeper should have some background information on the problem; others are mentioned in passing merely to inform the individual that the problems exist. Foulbrood was the common name given to one disease before the causal organisms were discovered. The term designates either one of two distinct diseases, American foulbrood (AFB) and European foulbrood (EFB). The common names have no specific significance in terms of origin or distribution.

The causal organism of AFB is *Bacillus larvae,* a spore-forming bacteria; EFB is caused by *Streptococcus pluton.* While AFB is a far more serious disease, either can be costly in terms of production loss and treatment time. Spores produced by *Bacillus larvae* are unusually resistant: they can withstand boiling water and lie dormant in equipment for 50 years. Both diseases impart a characteristic odor to the brood comb, hence the name "foulbrood." Even today inspectors use odor as a clue in diagnosing the disease. European foulbrood imparts a sour, yeasty odor to the comb; AFB has the characteristic odor of dead animal tissue. When food containing AFB spores is fed to a young larva, the spore germinates, multiplies, and kills it just before the cell is sealed. The only symptom at this time is a slight graying or dulling of the larva. Healthy larva are pearly white. The infected larva dies stretched lengthwise in the cell; it may have begun to pupate just prior to death. If the cell was capped, the cap may appear slightly sunken compared to a healthy one. As the larva continues to decay, it dries down on the lower side of the cell. As dead pupae dry down, a fragment of tissue sometimes remains attached to the top of the cell. This is usually described as a tongue — one field diagnostic feature. Larvae infected with EFB die about 4 days after hatching. Many are still coiled in the bottom of the cell, unlike those that died from AFB.

As the larva decays, it dries down on the lower side of the cell if infected with AFB; at the bottom of the cell, if infected with EFB. Bacteria rots the larval skin and turns the remains of the larva into a slimy mass of tissue. As moisture evaporates, the tissue becomes sticky. This is the basis of the ropiness test used in field diagnosis as another clue in distinguishing EFB from AFB. With a toothpick, matchstick, or similar instrument touch the remains. If AFB is present, the remains will be a rubbery mass, and no material will be drawn along when the stick is withdrawn. If EFB is present, on the other hand, the dead tissue will stretch out like a piece of dough or taffy. Old brood comb in storage is inspected for scales formed by dried tissue on the bottom of the cell. These scales are somewhat difficult to see unless the cell is held at the correct angle toward the light. EFB scales adhere loosely to the walls, while AFB scales are firmly attached. Dried-down AFB scales are uniform, while in EFB infection they are not uniform but resemble the position in which the larva died.

While the symptoms described here are fairly reliable diagnostic characteristics and an experienced beekeeper can be relatively certain of his conclusion, final proof is determined by microscopic examination. Some localities require laboratory confirmation before ordering colony treatment or destruction. Foulbrood is spread by man as well as bees. The bacteria are nonpathogenic to people and other animals, but organisms can remain alive in honey and/or pollen. There are differences of opinion on the likelihood of spreading the disease by feeding commercially pruchased honey from unknown sources. Exchanging combs of brood, honey, and pollen from a diseased colony to a healthy colony, however, is almost a guarantee of spreading the disease. In nature the disease is spread by robbing. As mentioned previously, bees are opportunists and will collect or steal honey from any source, especially a sick or weak neighbor. If this colony, which could be several miles away, is infected with foulbrood, robber bees will bring back the disease with the stolen honey.

It was because innocent beekeepers suffered losses that nearly every state passed disease-control laws of some type. Some states closely regulate the movement and importation of bees, and some actually prohibit importing bees or combs. There is widespread interest in developing or breeding strains of bees that are resistant to foulbrood. None have yet been found resistant to AFB, but some were claimed to be resistant to EFB. Some research directed toward this goal is being done in universities, much of which is related to nutrition. Some researchers refer to EFB as a stress disease.

At the turn of the century when foulbrood was quite prevalent and little was known about the disease, innovative beekeepers tried a number of techniques in an attempt to discover the magic cure. One technique that some found to work with EFB was as follows. The queens were removed from several diseased colonies, which were united into one strong colony. The developing queen cells were destroyed 9 days later, then requeened on the twentieth day with Italian stock. After 27 days the new queen was producing new, healthy brood. The rationale behind this treatment is that a strong colony does a much better job of housecleaning. Being queenless for a short time allows workers to remove debris, clean and polish cells, and reduce the infection to a level at which the reconstituted colony may survive and prosper. Other beekeepers modified this technique slightly but used the same basic idea. This procedure does not meet many legal requirements, nor was it recognized as a cure or treatment for foulbrood.

With the discovery of sulfa drugs and antibiotics in the 1940s and later bee-disease control was simplified. While they are legally approved for use by regulatory agencies, the drugs are a subject of controversy among beekeepers. Modern drugs are not a cure-all, but they do make life more pleasant for man and his livestock, including bees. Larval bees are susceptible to AFB only for the first 3 days of life. Newly hatched larvae can be protected with certain drugs (sodium sulfathiazole and oxytetracycline) during these critical days, and beyond this stage they are immune. Disease organisms on the surface of the comb or in the cell are covered either with wax or larval skin. After they are burned, they are no longer exposed to the susceptible larvae. If for some reason the bacteria was uncovered or exposed to a susceptible larva, the disease could recur. This point concerns some beekeepers and regulatory officials: the equipment is not free of disease. There is probably no simple, effective way to diagnose the disease or to warn the prospective buyer of used equipment that it has a history of disease. Equipment transported in interstate commerce has a similar problem.

Some beekeepers feel that relying on drugs rather than on burning creates the possibility that strains of bacteria might develop that would be resistant to currently available drugs. Drug-resistant microorganisms are indisputable biological facts, and the beekeeper should use drugs only when necessary and at the approved dosage or rate. A number of beekeepers object to the use of any drug, synthetic, or unnatural item around bees. While it is their legitimate right to insist on this for their own bees, it is also the right of others to utilize the latest technology, provided that it meets the legal requirements of local, state, and federal governments. The question of whether drug residues can find their way into honey has been researched by drug manufacturers as well as government laboratories, and laboratory procedures to detect residues in food are continually reviewed and updated in light of new technology. When drugs are used properly, no illegal residue will result, but beekeepers should use only approved drug treatments and not experiment with new ones. Testing new drugs is the duty of drug manufacturers, government laboratories, and state universities.

The techniques for treating colonies of bees with drugs are quite simple. They can be suspended in sugar syrup and fed to the colonies requiring treatment. Some beekeepers combine this treatment with that for nosema. Oxytetracycline is effective on both EFB and AFB. It can be mixed in powdered sugar and dusted over the bees and frames. Bees do not like to be covered with dust, so they will groom themselves and each other. Sufficient antibiotic is passed on to newly hatched larvae to protect them from the disease. The dusting technique is used when sugar syrup is not readily taken because of nectar flow. The advantage of a simple drug treatment for bee diseases, especially for a commercial beekeeping operation, is obvious. Diseased colonies can be quickly treated by nonprofessional labor. The cost is a small fraction of the value of the colony, which would otherwise be destroyed. Routine sanitary or cleanup procedures should never be replaced by drugs. It is good business for a beekeeper who discovers a heavily diseased colony with brood comb of questionable value to destroy it and sterilize items such as hive bodies, top covers, and bottom boards. From time to time interest is expressed in fumigating beekeeping equipment to rid it of diseases. The procedure, which includes controlled temperature and pressure, requires some engineering skill. One fumigant, ethylene oxide, is highly explosive if not handled properly.

Honeybee larvae can die from causes other than foulbrood. In some situations this can be serious, but most other causes of brood death are less severe to the colony. Low temperatures can kill brood. If a cold spell occurs in late winter or early spring when the queen is rapidly laying eggs and the cluster is not large enough to adequately cover the brood, some of them, especially around the perimeter, will die.

Wax Moths

Wax moths lay eggs in cracks and crevices between hive parts. After hatching the larvae tunnel through the wax and work toward the midrib of the comb, obtaining nourishment from impurities and debris in the comb. Wax is consumed but apparently not digested. Honey leaks from the openings, which affects the appearance and marketability of the sections. Wax worms feeding on foundation or in extracting supers rarely complete development. Once they are full-grown, the larvae spin dense cocoons among tunnels (and webbings) or attach to solid-wood parts (such as a frame or inner cover). The pupal stage varies from 8 to 60 days, depending on the temperature. The moth is about 3/4" long, with a wing spread of 1" to 1 1/4". It usually rests with its gray-brown wings folded up in a rooflike fashion. If the moth is disturbed, it runs rapidly; it is not an active flier.

As with most other pests, a strong colony is the best defense against wax worms. Worker bees aggressively attack and kill worms as well as moths if they can get to them. The greatest risk of wax-worm damage occurs when brood combs are put into storage. The hive bodies should be stacked on a solid floor (or on paper) in groups about

5' high. Cracks between the hive bodies should be taped, and a fumigant, paradichlorobenzene (PDB), put into each stack. About 6 tablespoons are needed approximately every 2 to 3 weeks. The stacks must be covered tightly so that gas does not escape — it kills larvae and adults but not eggs. Ethylene dibromide and carbon dioxide are also used as fumigants, but the engineering associated with their use has to be more precise. Two nonchemical treatments are also effective. Heat — 120 °F. (49°C.) for 40 minutes or 115°F. (46°C.) for 80 minutes — will kill all stages of the wax worm. Lower temperatures — 20°F. (6°C.) for 4 ½ hours, 10°F. (-12°C.) for 3 hours, or 5°F. (-15°C.) for 2 hours — will also kill these pests. In northern climates storing brood combs outside or in unheated buildings is commonly done without incurring wax-worm damage. Comb-honey producers keep section comb honey in a freezer. This eliminates the danger of wax-worm damage and preserves the flavor and quality of honey.

Insecticides

Bee deaths due to insecticides are not new. Records show that this problem started in 1870 with the use of Paris green, a pesticide used to control codling moths on apples. The problem intensified through the years for several reasons. The number of acres planted with crops that need protection from damaging insects has increased; specific chemicals that are highly toxic to insects have been developed; equipment such as aircraft and large ground sprayers can now apply insecticides to relatively large acreages during a short period of time; the elimination of roadside and wasteland weeds with herbicides reduces nonagricultural bee-foraging areas, so they have to work agricultural cropland subject to insecticide treatment. Mechanization may require additional insect-control sprays. For example, peas are now mechanically cut and shelled in the field, so any worms present on the vines will find their way into the finished product (a can of peas). Prior to mechanization peas were cut and hauled to a central location for shelling; any worms would be removed beforehand. Now a spray must be used prior to harvesting to assure the consumer of a worm-free product.

In some areas insecticide poisoning overshadows all other problems, including bee diseases; in others the problem is minor in terms of monetary loss to the beekeeper, but the conflict between beekeepers and insecticide users can become highly emotional. Insects are killed with poisons in three ways: fumigation, direct contact with the spray, or ingestion. Bees are no different from other insects in relation to insecticides: death is not unusual, nor should it go unsuspected because bees are insects. It is man's use or misuse of a chemical that causes problems. In some instances a conflict of interest develops. In some cases there is no reason to spray a crop if it is in full bloom and highly attractive to bees. For example, fruit trees bloom for a short period of time, and damaging insects can be controlled before or after blooming. In other cases the solution is not so simple: some plants bloom over a longer period of time, and the grower cannot indefinitely postpone insecticide treatment. Alfalfa and cotton are two such crops. A nontarget plant is sometimes unavoidably sprayed. Some weeds in a field of peas, for example, may be highly attractive to bees and sprayed in the process of treating peas.

Several points related to the insecticide-bee controversy should be understood by beekeepers as well as farmers. If an insecticide is applied to a specific field and it drifts or moves to another area not intended to be treated, the applicator has violated a federal law and is responsible for civil damages. If negligence is proven, criminal sanctions can be applied. The insecticide must be registered for the pest and for the crop. A farmer can legally spray his crop without regard to bees if he uses certain products. Some local governmental agencies may require farmers who intend to use certain products to obtain a permit before spraying. The permit may be denied if the crop is in a stage that is highly attractive to bees, or it may merely notify the beekeeper of the farmer's intention, and the beekeeper must then take appropriate action. In some areas beekeepers and farmers growing specific crops that require routine insecticide treatments agree to notify each other of their intentions. Beekeepers attempt to avoid placing bees near locations most likely to be sprayed, and farmers select a method of spraying least likely to injure bees. Needless to say, the controversy is far from being resolved. Bees are needed to pollinate over 50 important crops, and the farmers need to protect their crops from insect damage and destruction. As food becomes less abundant, the conflicts will intensify.

Chemicals are widely used in American agriculture. Among them are insecticides, herbicides, fungicides, growth regulators, hormones, fertilizers, and rodenticides. There are approximately 400 pesticides currently in use, about 20% of which are highly toxic to bees, 15% moderately toxic, and the majority (65%) essentially nontoxic at usual dosages. The most widely used chemical compounds for insect control are phosphates and carbamates, which are essentially nerve poisons. As would be expected, some are highly toxic to bees; others, moderately hazardous. If a highly toxic chemical comes in contact with a foraging bee, in all probability she will be killed and more than likely not return to the hive. If the hive is accidentally sprayed or if it is located alongside the treated field, a large number of bees will die in and around the hive. If field bees come in contact with highly toxic insecticide and no drift occurs in and around the hive, the effect on the colony would be minimal. Honey production would be lost for a period of time due to elimination of the field force. This field force would eventually be replaced, and the colony should return to normal. The typical problem that the beekeeper faces in this type of situation is more serious: one day his foraging bees are eliminated in field A; 2 days later the same thing happens in field B; and on the fifth day his bees are hit in field C. The net result is that he is seriously damaged.

Another far more serious type of poisoning occurs if pollen becomes contaminated with insecticide. One product known to contaminate pollen is carbaryl (Sevin), a carbamate insecticide that is relatively safe to birds and mammals because it does not readily penetrate the skin. Bees continue to forage for pollen in routine fashion even after the plants are sprayed. Carbaryl is relatively slow in killing insects and serves primarily as a stomach poison, so the bees unknowingly continue to carry insecticide-contaminated pollen to the hive. After the pollen is consumed by the hive workers, the carbaryl begins to act, killing workers who have eaten the contaminated pollen and sometimes even the queen and young brood. In serious insecticide-poisoning cases large numbers of dead bees accumulate around and inside the hive. Often the queen stops laying, and some brood will starve because there are not enough workers to care for them.

The insecticide problem is complex, and no simple solution appears to be in sight. In some situations proper timing of the insecticide spray helps to protect some bees, but some plants are attractive all day, such as alfalfa and cotton. This gives the grower little choice, other than night spraying, if the crop is to be protected from plant-damaging insects. Some plants are attractive only in the morning — corn is one example. Only the ear of the corn must be protected from worm damage. Bee losses have been reduced when the insecticide was applied in the late afternoon and evening, using equipment that selectively treats only the ears, leaving the tassels unsprayed. A number of serious honeybee kills resulted when corn was treated in the morning by airplane. The entire plant was sprayed while bees were foraging for corn pollen. Corn is not highly attractive to bees: they will collect pollen from it only if it is not available from other plants. Colonies can be covered with burlap or dark plaster for 1 to 2 hours during insecticide treatment — up to 2 days if the burlap is kept wet. In most cases bees can reenter a treated field in 2 or 3 days without significant injury — sooner with some very short residual materials.

Some people have expressed concern over the possibility of honey contamination. Fortunately, this has not occurred to date. The bee has several built-in safeguards in protecting its food supply. Nectar is normally not exposed to most insecticides. If it does somehow become contaminated, the bee has further safeguards. She will probably not return to the hive with poisoned nectar. If she is not killed immediately, there is a good possibility that she will get lost and not find her way home. If she does find her way to the hive and she is contaminated with some peculiar odor, guard bees may not let her in. If she happens to get in, the content of the crop is fed to other bees in routine regurgitation. Some moisture is lost, so if a residue of some type were present, it would be concentrated and eventually kill one of the receiving workers before it is stored. These safeguards explain the fact that no pesticide residue has been found in honey.

The African Killer Bee

The Africanized honeybee has flourished in Brazil for nearly 20 years, but the publicity given this creature in the last few years by news media makes it appear a very recent event. Professional beekeepers are well aware of its desirable and unfavorable characteristics.

Productivity of the African bee (*Apis mellifera adansonii*) is well documented. It prompted the government of Brazil to send Dr. W. E. Kerr to Africa in 1956 to select some queens for crossing with native stock (which were primarily of Italian stock). The basic purpose was to increase Brazilian honey production. Of the 133 queens brought back, 47 survived, 1 from Tanzania and 46 from Pretoria, South Africa. They were established in Sao Paulo in November 1956. From this group Kerr selected 35 that he thought most promising and transferred them to a eucalyptus forest. He intended to obtain good hybrids through the use of artificial or instrument insemination. Hybrids would then be used for further research and breeding, and high-quality stock would be released to beekeepers. Dr. Kerr took the necessary precaution of confining them in escape-proof hives with double queen excluders over the entrances. A visitor saw the hives and noticed piles of pollen in front of the entrance. He thought he would do Dr. Kerr a favor by removing them. Before Dr. Kerr discovered the tragic error, 26 colonies had swarmed. In addition to his highly defensive nature the African bee has a tendency to swarm frequently.

It spread rapidly throughout Brazil. By 1975 it was estimated that the African bee had hybridized with over 67% of all Brazilian bees.

Estimates differ as to the actual rate of dispersal. Some authorities admit that the reported spread of several hundred miles per year is open to question. It is believed that more enterprising and progressive beekeepers in distant parts of Brazil could have helped the situation by obtaining wild swarms with new blood for their own apiaries.

Study teams from the United States and other countries traveled to Brazil to see the problem firsthand; news reporters did the same. As would be expected, widely differing reports have been published. Journalists coined the name "killer bee" to identify the African-Brazilian hybrid. In summary, the teams found that the hydbrid that moved south to more temperate climates where beekeeping and agriculture are more highly developed was apparently quite different from the one that spread north to the tropics. In southern Brazil a number of hobby beekeepers went out of business. It is reported that many of them were more interested in keeping bees than in producing honey. The more progressive commercial beekeepers, using standardized Langstroth hives, now claim to be producing up to twice as much honey than with Italian bees. Like most beekeepers in the United States, they continue to selectively eliminate and destroy difficult-to-handle colonies. In northern Brazil beekeeping is more primitive in terms of management and available equipment. There are more natural sites where a swarm can become established. Some swarms settling in villages and metropolitan areas created havoc. Some incidents were overly dramatized; others were not. Insofar as honey production is concerned, facts speak for themselves. Before 1970 essentially no honey was produced or exported from Recife in the state of Pernambuco. In 1972 800 tons were exported to Liverpool, England.

How productive are the African bees in Africa? One of Kerrs' queens came from Tanzania. That country's production, with what is considered primitive equipment, produces about 600 tons of wax and 10,000 tons of honey per year. Production in other countries, such as Angola, South Africa, and Rhodesia, is similar. In the two latter countries, where modern frame equipment is used, the bee is manageable. In order to reduce the highly defensive nature of the Africanized Brazilian, as some now call this bee, the Brazilian government started a program of distributing virgin Italian queens, which would naturally mate with the wild drones. The results appear quite promising; productivity remains high; and the resulting colonies are easier to handle. Paraguay has tried to introduce beekeeping to the

Indians by supplying them with European stock from Canada. Within the last few years there have been reports that honey, now collected from wild colonies, is available on local market.

Why does the African bee produce more honey? Observations in Brazil suggest that about 30% more of the bees are active during the first hour of the morning compared to the native Italian stock; the same is true at the end of the day, especially for the last 25 minutes. Aggressiveness has been measured using several behavioral characteristics. Interesting differences are noted between the African, the Brazilian-African hybrid, and the Italian. The African bee stings at an average of 160 yards from the hive; Brazilian-African hybrids, 38 yards; pure Italians, 21 yards. When hives were intentionally disturbed, the African bee became agitated in 22 seconds; the Brazilian-African hybrids in 30 seconds, and Italians in 43 seconds. Once disturbed the African bees were still not calm after 28 minutes. They are nervous and quickly make the decision to defend the colony.

CHAPTER SEVEN

Watch Bees Work

Building and Operating an Observation Hive

Observation hives provide interesting, educational, and entertaining exhibits in classrooms, parks, zoos, campgrounds, places of business and homes. They have been used for many years to study the behavior and activities of bees. Bees work within an observation hive as industriously as they do in their "natural" or normal environment.

Basically, the observation hive is simply a miniature transparent walled hive. While they can be constructed in any shape or size, most observation hives are designed to hold one or several standard combs. The most satisfactory designs are those that are one comb wide, and several (3 to 6) combs high. The reasons are simple. The total amount of pollen and honey stored is visible. The queen will always be exposed so she can be observed at any time. More workers can be observed performing their routine hive duties.

While the height of the observation hive is optional, bees confined to a single comb will have difficulty accumulating sufficient honey and pollen to carry them through periods of rainy and cold weather or when no nectar is available. In this case they will have to be fed. A strong colony will have a large percentage of its workers collecting nectar or honey, thus providing a more interesting and active exhibit.

An observation hive can be constructed by simply supporting two parallel sheets of glass or acrylic plastic with suitable framing that has an entrance-exit opening to the outdoors. A comb is suspended within this chamber. If no comb or foundation is provided, bees will construct their own.

If this is your first experience working with bees, you might want to ask a beekeeper to help you get started. Hobbyst beekeepers are generally willing to share their experiences and offer suggestions or even assist you in getting started.

You can purchase a queen and bees from commercial sources. It is much simpler, however, if you obtain a marked queen and combs containing brood and honey from a beekeeper. Both methods of starting an observation beehive are described.

Construction and Location

Anyone with a home workshop can make the observation beehive shown in figure A. Use acrylic plastic (Plexiglass or Lucite) rather than glass because it will not break. Plans are drawn to use the standard 6 ¼-inch shallow frame; other size frames will need appropriate adjustments.

While frames can be made in a shop, you can purchase them, along with an appropriate foundation, from a bee supply house.

It is very important to use well-seasoned wood and make construction exact in order to avoid openings for the bees. The frame holders (Figure E) should be interchangeable, so they must be identical. The illustration suggests the hive rest on a pipe (Figure B). This allows flexibility, so the hive turns on the threads. Provide a stop so the union does not separate.

The height is adjustable to suit the needs of the audience or viewers. Locate the outside opening so it won't interfere with people, especially small children. Also avoid open doors, windows, or sidewalks. A second story window or higher is ideal for the outlet, but not always possible or necessary. Do not put the hive in direct sunlight. This will cause overheating and probably will kill the bees.

Install the observation hive in the spring, as soon as daytime temperatures go above 60°F. It is best to install it, however, when fruit trees or dandelions begin to bloom. If bees are installed early, be sure to have enough food (sugar or honey) in reserve so that they don't starve on cold days.

Installation from an Existing Colony

Three combs of brood and honey are about the optimum amount needed to start the colony. One comb should contain substantial amounts of sealed brood. When starting the hive early, use two combs containing honey—one if later. The third comb can either contain eggs or developing larvae. Include sufficient bees so they can adequately tend the brood and queen. Be sure the queen is laying and conspicuously marked so she is readily identified by the observers.

Place the combs with bees, brood or honey in a frame-holder (Figure E) and cover them with sheet metal slides (9-Figure A). Use masking tape to securely fasten the metal slide covers to the frame holder. Normally bees can be maintained in this frame holder for two hours. For longer time periods, use a screen in place of metal slides on one side; otherwise the bees might suffocate. Be sure that this screen is firmly attached to the frame holder with tape or by some other means. Do not place the bees in direct sunlight while holding or transporting.

Installation

First, locate the observation hive exactly where you want it and check that the opening to the outside is not blocked.

Place the comb in a frame holder with honey in position (7-Figure A). Then place another comb in a frame holder with sealed brood in position 6 and the frame containing eggs in position 5. While these positions are not absolutely essential, the bees tend to move the honey from position 7 upward. This will open cells for the queen to lay eggs. Keep the queen and brood in the lower portions of the observation hive. Normally honey is stored in the top of the hive. Place either comb or foundation in positions 4, 3 and 2, and as the population increases, bees will either store honey or the queen will lay eggs in this area.

With the frame holders containing bees in position, the sheet metal slides can be removed. Bees can then mix and spread throughout the observation hive. Secure the end cover (1-Figure A) in place.

Other Suggestions

Masking tape is very handy to have available. You can use it to secure the metal slides (9-Figure A) or in aligning and holding in line the holders in positions 1 to 7 until the end is secured. If these holders are not in line, bees will escape.

Starting from Packaged Bees

If you start from packaged bees, purchase a queen and about two pounds of bees. Do not attempt this unless you are willing to perform

SIX FRAME OBSERVATION BEE HIVE

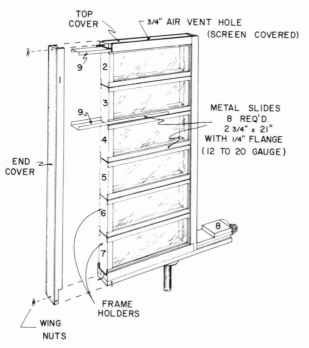

TOP COVER

3/4" AIR VENT HOLE (SCREEN COVERED)

9

2

3

9

4

METAL SLIDES
8 REQ'D.
2 3/4" x 21"
WITH 1/4" FLANGE
(12 TO 20 GAUGE)

END COVER

5

6

8

7

FRAME HOLDERS

WING NUTS

PERSPECTIVE (FIG. A)

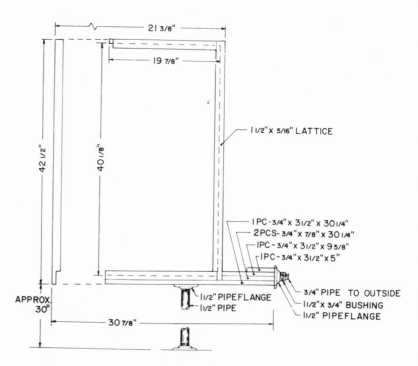

21 3/8"

19 7/8"

1 1/2" x 5/16" LATTICE

42 1/2"

40 1/8"

1 PC - 3/4" x 3 1/2" x 30 1/4"
2 PCS - 3/4" x 7/8" x 30 1/4"
1 PC - 3/4" x 3 1/2" x 9 5/8"
1 PC - 3/4" x 3 1/2" x 5"

3/4" PIPE TO OUTSIDE
1 1/2" x 3/4" BUSHING
1 1/2" PIPEFLANGE

APPROX. 30"

1 1/2" PIPEFLANGE
1 1/2" PIPE

30 7/8"

SIDE VIEW (FIG. B)

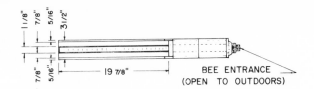

TOP VIEW OF HIVE BASE (FIG. C)

REMOVABLE METAL
CLEANOUT TRAY
1" W x 3/4" H

CONSTRUCTION DETAIL
BEE PASSAGE-WAY (FIG. D)

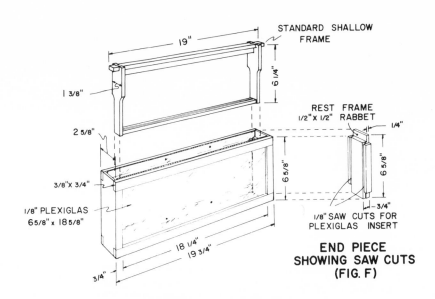

STANDARD SHALLOW
FRAME

REST FRAME
1/2" x 1/2" RABBET

1/8" PLEXIGLAS
6 5/8" x 18 5/8"

1/8" SAW CUTS FOR
PLEXIGLAS INSERT

END PIECE
SHOWING SAW CUTS
(FIG. F)

PERSPECTIVE OF FRAME 8
FRAME HOLDER (FIG.E) 8 REQ'D.

extra work or have had some experience working with bees. Be sure to study the techniques and procedures used in introducing packaged bees.

Procedure

Shake the bees into one or two frame holders containing combs and introduce the queen. Either feed the bees sugar syrup and pollen supplement or have honey available.

If no drawn comb is available, provide a feeding device, such as an inverted bottle with sugar syrup, or a division board feeder. This must fit inside one of the frame holders. If you start with packaged bees it is essential to provide supplemental food, because bees will not be able to accumulate enough stores to carry them over through rainy or cold days in the spring, which may occur at any time.

Maintenance During Summer and Fall

Many variables determine the amount of honey stored in the summer. If the hive becomes too crowded, remove one frame and destroy or place the bees in another colony at least a half-mile away. If the colony becomes small, add additional bees or, if necessary, replace the queen. Long periods of drought or continuously wet, rainy weather result in honey depletion. In this case, feed sugar syrup.

As mentioned, the unique feature of the hive in Figure A is that you can manipulate it indoors without dismantling. If done properly, no bees will escape. Suppose you wish to remove the bees and frame in position 5 and replace it with an empty frame. The first step is to insert the metal slide (12 to 20 gauge) between positions 5 and 6 (9-Figure A). The metal slide should close the TOP of the frame holder in position 6. Secure this with masking tape. Next, insert a second metal slide between positions 5 and 6, covering the BOTTOM of the frame holder in position 5. Secure this with masking tape.

Take the fourth metal slide and insert it between 4 and 5, closing the BOTTOM of the frame holder in position 4. This confines the bees that are in frame holders in positions 2, 3 and 4 into that area. The bees are also confined exclusively to the frame holder in position 5, and into position 6 and 7. All that is now necessary is to slide out the frame holder in position 5.

It is quite important that you have someone assist you, or the frame holders in position 2, 3, and 4 will drop out of place. You must retain the frame holders in place so you can insert a new or empty frame holder into position 5.

Once installed and in operation, you can perform a number of observations, manipulations and simple experiments with your hive.

If you want an end view of comb, cut several drawn combs into sections and install them perpendicular to the length of frame holder. Space the sections about ½ to ¾ inch apart. In some cells developing larvae may be visible through the transparent side. You might space one or two sections several inches apart. Observe what bees do with that extra space.

References

BECK, B.F. and SMEDLEY, D. 1971. *Honey and Your Health*. Bantam Books. New York.

CRANE E.E. (Ed.). 1975 *Honey: A Comprehensive Survey*. In cooperation with Bee Research Association. Russak, N.Y.

EBLING, W. 1975. *Urban Entomology*. Univ. of Calif. Div. of Agr. Sciences, Berkeley, Ca.

GOJMERAC, W.L. 1975. *Honey: Guidelines for Efficient Production*. Univ. of Wis. Ext. A 2083. Madison, Wi.

JAYCOX, E.R. 1969. *Beekeeping in Illinois*. Univ. of Illinois Cir. 1000. Urbana, Il.

LANGSTROTH, L.L. (Ed. Grout). 1975. *The Hive and the Honeybee*. Dadant and Co. Hamilton, Il.

MACKENSON, O., TUCKER, K.W. 1970 *Instrumental Insemination of Queen Bees*. Agr. Handbook 390, ARS, USDA.

MOELLER, F.E. 1976. *Two Queen System of Honey Bee Colony Management*. P.R.R. No. 161, ARS, USDA, in coop. with Univ. of Wis., Madison, Wi.

MORSE, R.A. 1975. *Bees and Beekeeping*. Cornell Univ. Press. Ithaca, N.Y.

Index